BREAKING the MOLD

Changing the Face of Climate Science

DANA ALISON LEVY

books for a better earth™

holiday house • new york

To Cliff, Alex, Anita, Rae, Rocío, Lila, Marshall, Daniel, Devyani, Gabriela, Chris, Katharine, Valerie, Caroline, Rupert, and Kelly...for all you do, and all you gave to this project, I am so grateful.

A **Books for a Better Earth**™ Title

The Books for a Better Earth™ collection is designed to inspire young people to become active, knowledgeable participants in caring for the planet they live on. Focusing on solutions to climate change challenges, the collection looks at how scientists, activists, and young leaders are working to safeguard Earth's future.

Copyright © 2023 by Dana Alison Levy
All Rights Reserved
HOLIDAY HOUSE is registered in the U.S. Patent and Trademark Office.
Printed and bound in October 2022 at Toppan Leefung, DongGuan, China.
This book was printed on FSC®-certified text paper.
www.holidayhouse.com
First Edition
1 3 5 7 9 10 8 6 4 2

Library of Congress Cataloging-in-Publication Data

Names: Levy, Dana Alison, author.
Title: Breaking the mold : changing the face of climate science / Dana Alison Levy.
Description: First edition. | New York : Holiday House, [2023] | Includes bibliographical references. | Audience: Ages 8–12 | Audience: Grades 4–6
Summary: "Sixteen profiles of scientists who are changing the face of science and the future of Earth through their research"–Provided by publisher.
Identifiers: LCCN 2022001753 | ISBN 9780823449712 (hardcover)
Subjects: LCSH: Minority scientists–United States–Biography–Juvenile literature. | Minorities in science–United States–Juvenile literature.
Classification: LCC Q141 .L352 2023 | DDC 509.2/2–dc23/eng20220521
LC record available at https://lccn.loc.gov/2022001753

ISBN: 978-0-8234-4971-2 (hardcover)

TABLE OF CONTENTS

The greatest threat our species has ever faced is climate change and the way that it is rapidly changing our planet.

FOREWORD

CLOSE YOUR EYES AND PICTURE A SCIENTIST.

Was it a white dude in a lab coat, surrounded by test tubes and beakers?

If so, you're not alone. In a scientific study (yes, those scientists research everything!) researchers asked children ages five to eighteen to draw pictures of scientists, and looked at the results across fifty years. In the 1960s and 1970s, less than 1 percent of students drew their scientists as female. By 2016, however, approximately 34 percent of students' drawings showed women. And girls? They drew more than half of their scientists as women! In some ways that's not surprising—the number of women with jobs in science rose steadily throughout those five decades, though there are many fields, like physics and astronomy, where women still make up only a small percentage of the whole.

But gender is not the only area where diversity in science is important. The stereotype of the white, able-bodied man leaves out many other practicing scientists as well. Disabled scientists, Black and Indigenous and Latinx scientists, scientists with learning differences, and countless others exist in science who don't find themselves in the stereotype.

And make no mistake, diversity among scientists leads to better science. The greatest threat our species has ever faced is climate change and the way that it is rapidly changing our planet. Weather patterns are more extreme, so that droughts and wildfires are increasing in one area while floods and sea-level rise are happening elsewhere. All around the world, the communities and people who are most affected by climate change are often some of the most oppressed. In order to come up with solutions that will truly help all of humanity, we need all of humanity to be a part of the process.

One area where non-white scientists are far too rare is environmental science. The vast majority of environmental-science majors at colleges and universities are white, and the staff and board members of environmental organizations reflect the same trend. But recently there has been a renewed focus on the intersection of racial and social injustice and climate change. These connected issues are often discussed using terms like "environmental justice" and "environmental racism," which refer to the way systems of racism and other forms of oppression make climate change, pollution, and other environmental risks worse for certain communities. In other words, richer, whiter communities tend to be less impacted by environmental problems and climate change than poorer communities and people of the global majority, a term that acknowledges the majority population of Black, brown, and Indigenous people around the world. But when the scientists and policy makers all belong to the more privileged groups, the decisions they make often continue to prioritize their own interests!

With new awareness comes new energy, and there are countless scientists working on everything from atmospheric science (studying weather patterns) to ocean health to plant resilience. There are Black scientists advising the president of the United States, and disabled scientists doing fieldwork. There are scientists who started families before going back to school, and scientists working to make sure that LGBTQ+ students and faculty at their schools feel welcome to be their whole selves in the lab and elsewhere.

The picture is changing, for sure. Diverse scientists of all genders are in laboratories, on research vessels, in classrooms, and knee-deep in mud doing fieldwork. They are breaking the mold, smashing the stereotype of a certain kind of person in a certain kind of setting, and in doing so they are redefining what it means to look like a scientist.

Diverse scientists bring different and valuable perspectives to their work, and force the scientific community to question long-standing assumptions and biases that were always considered "normal" just because they were normal for the small group of people who were in charge. These discussions and disagreements aren't always easy, but science isn't supposed to look for the easiest answers, and scientists should welcome the opportunity to learn more and even realize they were wrong. After all, that's how great discoveries are made!

In these pages you will read about scientists who felt, again and again, that they did not fit the mold of what a scientist should look like. Some of them are well-known in their field, and some of them are just starting out. Some of them loved their journey to becoming a scientist, while others struggled and admit that they are still not sure they belong. You will read about people like Kelly Luis, who uses satellite images to study ocean color and health, or Daniel Palacios, who was so fascinated by whales and marine life that he hitched a ride on a research vessel to the Galápagos by offering to scrape the hull and paint trim. You'll meet scientists from Atlanta, Georgia, and Cape Town, South Africa.

None of them have the same story. But all of them share the same truth: even if you feel like you don't fit the picture of a scientist, that doesn't mean you're wrong.

It means the picture is.

Because scientifically speaking, if you like asking questions, if you're curious about the world, if you're committed to helping be part of the solution for our changing planet, if you're willing to work and try and fail and try again . . . well, look in the mirror. You definitely look like a scientist.

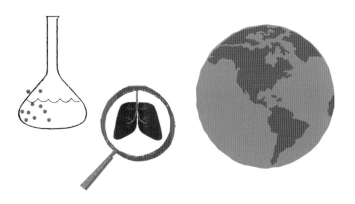

"I love my home, and I always wanted to find more ways to share about this place, through my cultural perspective. And science became a tool for that."
—Cliff Kapono

CLIFF KAPONO

Cliff Kapono might be in a university laboratory, running complicated computer tests to understand the health of our oceans. Or he might be halfway around the globe at an international surfing competition, with photographers lining up to see him ride some of the gnarliest waves on the planet. He is a professional surfer and a professional scientist, an identity he wears with pride, and he is confident that both of these careers help him be a better advocate for the planet. But getting to this point wasn't always easy.

As a child on the Big Island of Hawaii, Cliff didn't see much difference between science and just being curious about the world around him. Growing up in an **Indigenous** Hawaiian household, learning about the land and the natural world was a part of his culture and his family upbringing.

◄ Cliff Kapono catches a wave.

Often we come across references to people, animals, or plants that are **indigenous** to a place. Indigenous is another way of saying native or naturally occurring in a geographic location. So Indigenous Hawaiians are the people who lived on that land before European explorers colonized the islands in the eighteenth century. And Indigenous people, more generally, refers to the Native communities that existed before settlers and colonizers.

In the United States, Indigenous people are sometimes called Native Americans, or American Indians, or ideally by the specific name of their tribe. (Did you know that there are currently 574 federally recognized tribes in the United States?) While there were many more tribes before colonizers arrived and killed off local populations through war and disease, enrolled tribal members and other Indigenous people of the land we call the United States are still very much part of our current culture!

Similarly, learning to surf was an obvious skill to have, living in the birthplace of surfing and the center of the global surf community. When he was young, Cliff dreamed of becoming a professional surfer, or a professional scientist. But he felt both options were unlikely. He says, "Sometimes coming from a **marginalized** community, the dream of being a professional is met with some harsh statistics that can be discouraging. Those statistics can keep you from pursuing something you really love."

But Cliff kept at it. In his home, science was celebrated and integrated into Hawaiian culture, but at school, "I was kind of a rascal, getting in trouble with friends. Then one day at an assembly, we were making noise and goofing around when they announced the Bill Nye the Science Guy Award, and they gave it to me! I was number one in science at the school! Something changed that day."

When Cliff started taking more challenging classes in high school, it didn't always go well. He says he often struggled in those classes, getting Cs and Ds. This could have discouraged him from sticking with science, but, he says, "Something in me believed that even though I wasn't really good at this from the objective perspective, that was okay. I knew I really enjoyed it. I loved that when

"I loved that when I looked at a tree I was able to speak about it as a Native Hawaiian, but also that I understood it on a microscopic level."

I looked at a tree I was able to speak about it as a Native Hawaiian, but also that I understood it on a microscopic level. It demystified a lot of the knowledge passed down in my culture. I could talk about what was happening with the trees through stories of gods and goddesses, or through xylem and phloem. I had a choice."

Cliff says that in a way, not getting good grades took the pressure off. Because he wasn't so great at school, he never felt he had to sacrifice his Native culture. He felt a bit like a rebel, like he was going to practice science and learn more despite not looking like a typical scientist. At this time he was also at a boarding school that forbade him from surfing, so he was sneaking out to surf in secret. "For me science *and* surfing were anti-establishment, a way for me to be cool. Like a graffiti artist or someone playing in a garage band, I thought of my science as a way of being socially relevant and rebellious."

He continues, "The more I got into science, the more it became about

The word **marginalized** comes from the idea of something being pushed to the sides, or the margins, as opposed to being able to take up space in the center of things. Marginalization is a way of treating a group, or individual, or even an idea as unimportant or insignificant.

While we don't mean that people are literally pushed to the side when we use the expression "marginalized," we do mean that their importance is pushed to the side! Often people from marginalized communities are underrepresented in government policies, or discriminated against in employment, housing, or education. Because of this they have to work harder than those in wealthier or more powerful communities to achieve the same goals.

understanding how it fit into larger society. I love my home, and I always wanted to find more ways to share about this place, through my cultural perspective. And science became a tool for that. But I continued to be a secret surfer. I didn't really let the world know I surfed, because I didn't think people in science were willing to celebrate these two separate professions."

While he was at college he continued surfing, and would take as many surf trips as time and money allowed. But when an injury kept him out of the water for a little while, his grades really improved. For the first time, he saw the payoff for studying hard. He also found a particular new skill: as he was learning all kinds of cool things about the world through science, he was also learning how to communicate those ideas to friends and colleagues in the surf community.

Cliff was getting paid to travel as a professional surfer, as well as pursuing his education in science, but he wasn't combining them. It wasn't until a graduate-school professor recognized Cliff's growing popularity as a surfer that he thought there might be a way to use his surf career to further his scientific research. His professor thought Cliff's unique position would allow him to create powerful connections between ocean scientists and ocean users. And that's when they designed the Surfer Biome Project. Cliff says, "But my professor was clear: this was an experiment, and if I failed, I might not graduate."

The Surfer Biome Project had Cliff doing a lot of what he had always done: traveling the world to places where huge waves attracted famous surfers. But this time his goal was a little different. Instead of grabbing his board and diving in, he was grabbing . . . cotton swabs! The goal of the project was to try to learn whether people who spent a lot of time in the ocean—like surfers—had different **microbes** in their bodies than other folks.

There are a lot of reasons to want to better understand how microbes from the ocean impact humans, and how humans are impacting the oceans.

One example is how pollution on land flows into rivers, and ultimately into the ocean, and can be absorbed in the bodies of people who swim or surf a lot.

One issue that Cliff and the Surfer Biome Project studied was the role that ocean pollution plays in how people respond to important lifesaving drugs like antibiotics. All around the world, wastewater—from our toilets, showers, and sinks—is treated in processing plants and flushed out to sea. And of course rain causes flooding that washes still more untreated water into the ocean. The wastewater treatment plants are supposed to ensure that potentially dangerous chemicals and microbes are filtered out. But if traces of antibiotic medications are found in wastewater, and those traces enter the ocean, can they also enter the **microbiomes** of surfers and marine life? The project aims to understand whether this is happening, and to track the impact on surfers and their microbiomes.

To answer this, Cliff traveled through the waves of Ireland, Morocco, Chile, and beyond, taking samples from over five hundred surfers. He got samples by asking them to rub cotton swabs on different parts of their bodies, including their mouths. He also asked if people would volunteer something

WHAT IS A MICROBE? WHAT IS A MICROBIOME?

Microbes are tiny organisms, invisible except under a microscope, that are absolutely everywhere. There are different microbes in the oceans, lakes, soil, and even our bodies. Some are helpful, like the ones that live in our gut and help us digest food. Some are harmless. Some are . . . not so helpful, and can make us sick. But love them or hate them, there's no avoiding them!

The collection of microbes in a specific environment, whether as small as your stomach or as large as the ocean, is called a **microbiome**. Microbiomes are important because your microbiome helps all the different microbes in your body work together to keep your systems running smoothly. And different people, different environments, even different parts of the ocean all have different microbiomes. So learning about the microbes that are hiding in plain sight can tell scientists different things about how humans and their environments interact.

WHAT IS AN ANALYTICAL CHEMIST?

Despite loving the ocean and spending much of his career working on protecting the planet, Cliff did not major in marine biology or environmental science, two popular areas of study. Instead he got his degrees in analytical chemistry, which is essentially the work of studying and analyzing the chemical compounds that make up matter, which is the word for all the stuff on Earth! Analytical chemists use computers, math, and lab equipment to figure out literally thousands of different problems, from how to ensure food and medicines are safe to understanding human impact on the environment.

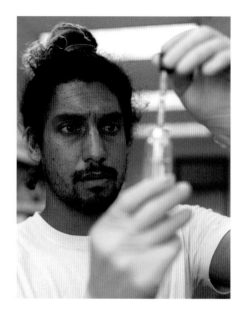

Cliff in the lab

else: poop! Believe it or not, fecal matter holds a lot of answers about the different microbes living in our bodies.

Cliff's experiment didn't fail, and that project cemented his identity as a surfer, a scientist, and an environmental steward. He feels that his success as a surfer gives him a bigger microphone to talk about science and the importance of conserving our oceans.

For Cliff, the science he studied has provided a tool kit he can use in many different ways. As an **analytical chemist**, he looks at the world by studying the chemical and molecular profiles of things. He says, "I've learned that my

WEIRDEST PART OF THE JOB:

"The weirdest part for me was the Surfer Biome Project in graduate school. I was looking into the microbiome of people's guts, which means their gut bacteria. You can learn really interesting things about the health of the environment by looking at what's in our guts. But since I couldn't actually look into the guts of all these people, I needed a proxy. And the best proxy is a stool sample. Yeah, that means poop. So I would have to ask the surfers, 'Can I have some poop sample?' People definitely look at me funny."

"Telling scientific stories and sharing them widely, helping the broader community understand why the science is important, is a critical part of my job."

passion is not on a lab bench. But I do love looking at data and finding trends that tell a story." He stresses that it's important to be honest with yourself about what you enjoy, because you will spend a lot of time doing it! For Cliff, one thing he realizes he enjoys is using photos and video to tell his stories. While on the surface those movies don't have anything to do with his formal science training, he says, "Telling scientific stories and sharing them widely, helping the broader community understand why the science is important, is a critical part of my job."

Cliff is quick to stress that now, as a professional surfer and scientist, he is able to share both his passions, but that wasn't always the case. "I thought about quitting a lot," he says. "I had a professor tell me that I had to quit surfing if I wanted to work in his lab, and I was pretty discouraged. I told my dad I was thinking about quitting, not because I needed his approval but because I valued his opinion. And it was hard, because I was one of the first in my family to have this kind of success, to get awards and be recognized in this conventional Western science setting. So I knew he would be disappointed, but he just said, 'You'll make the right decision, whatever you choose to do.' And that helped me calm down and rethink my decision. And I decided I was going to finish what I started."

"Failure doesn't mean you don't have power! Failure empowers you to figure out other paths, and to recognize how badly you want something."

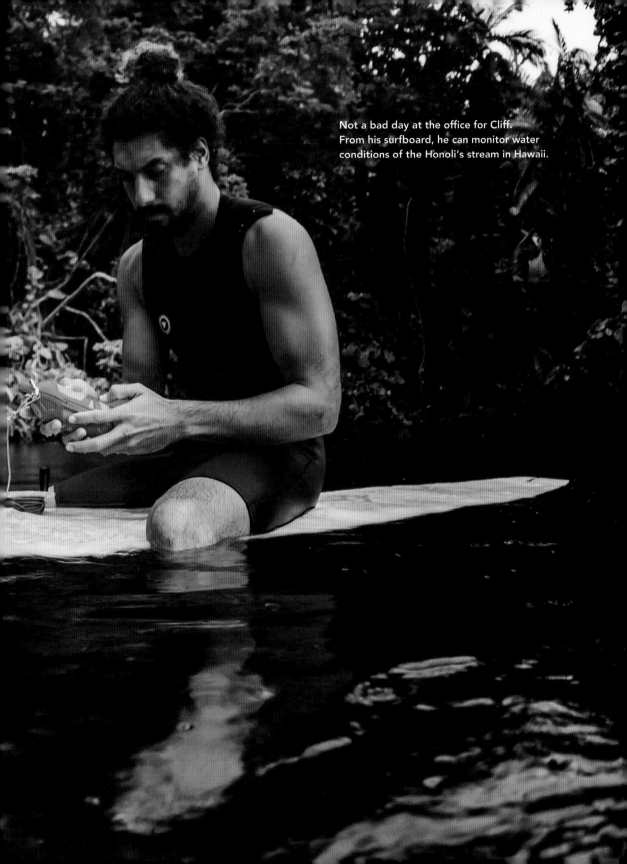

Not a bad day at the office for Cliff. From his surfboard, he can monitor water conditions of the Honoli's stream in Hawaii.

For Cliff, nothing takes the place of immersing himself underwater and studying what he finds there.

"We all need islander thinking, so we can all be better environmental stewards and better global citizens."

Cliff recognizes that it isn't always easy. "There is still so much trauma that exists from my path in academia, and that stems in part from the intergenerational trauma of what happened to Indigenous people in Hawaii. But for kids who are in a similar position, whether Hawaiian or not, I would urge them to recognize their power. Failure doesn't mean you don't have power! Failure empowers you to figure out other paths, and to recognize how badly you want something. So it comes down to the question: How badly do you want it? Only you can answer that."

Furthermore, Cliff points out, a scientist's job is to become an expert in whatever interests them, and learn how to think about the world in unique ways. He says, "As scientists we get rewarded to think about how we can improve our planet. It's one of the best jobs ever, and you don't need to have a certain degree to do it. You just have to wake up every day and think about what problem you're going to help fix. And realize, if I don't fix it today, that's okay, because I'll keep working on fixing it tomorrow. There's no pressure, because we just stay with the work, and that's how problems get solved."

Cliff sees storytelling as a natural extension of his work as both a surfer and a scientist. "I like to make it clear that not everything is black and white. Ultimately we, collectively, have to look at some big decisions about the planet. So what's the solution? I think it's about what we call 'islander thinking'—because when you live on an island, you know you're part of a collective. So we all need islander thinking, so we can all be better environmental stewards and better global citizens."

"I became more and more curious, and more and more excited about the things I was learning, and that kept pushing me forward." —Alex Moore

ALEX MOORE

Alex Moore has a job in science that takes them from downtown New York City, where they work at the American Museum of Natural History and Columbia University, to mangrove swamps in American Samoa, where they study ecology and coastal ecosystems. They're an educator who works with high school students up through graduate students, and a researcher who might be in the field for weeks or presenting at conferences around the world.

Alex recognizes that trying to separate science from elements of identity, such as race or gender expression, is one of the great lies scientists used to tell. While science should be objective, every scientist brings their own biases, personal experiences, and cultural norms to their work. As a **nonbinary**, queer, Black person, Alex feels strongly that they need to share all of themself

◀ Traveling to places like Hawaii, where they are in this photo, is one of the things Alex loves about their work.

WHAT DOES NONBINARY MEAN?

Nonbinary refers to a person who has a gender identity that isn't exclusively categorized as a woman or a man. For a long time many people—including scientists!—assumed that there were only two biological sexes, male and female, and that a person's gender was automatically matched to whichever of those labels they were assigned at birth. But gender and biology are both more complicated than that (for instance, on the biology side, some people are born with a combination of male and female reproductive organs, which is referred to as intersex, and is separate from the gender they identify with as they grow up).

A nonbinary person might feel that they don't fit neatly into a box that defines them as a boy or a girl. They might understand that they are a mix of characteristics that are often attributed to either males or females. In fact, nonbinary people often want to break away from the assumption that there is a specific set of boy traits or girl traits at all. Instead of two labels, they hope to allow people to express themselves in whatever way fits them best. Often nonbinary folks, like Alex, use the pronoun "they" instead of "he" or "she," and sometimes they will use the prefix "Mx." (pronounced *mix*) instead of "Mr." or "Ms."

Some of you might already use the label "nonbinary" yourselves or know folks who do. Others of you might be learning this for the first time. But science backs it up! Both gender and sexual biology are far more complex than two distinct buckets. All throughout the natural world, from mammals (like humans) to reptiles to plants, the way living things are identified, reproduce, and work in community is not as simple as just male and female. Human society is catching up to that. It's important to be respectful of people's identities and to use the names and pronouns that they want, so that everyone is comfortable being their most authentic selves.

When Alex was a kid they watched a ton of Discovery Channel and National Geographic and were always asking questions about everything.

Alex as a young graduate

when talking to students, other scientists, and the larger community. While it might cause discomfort, it also opens dialogues and raises awareness that everyone belongs in science. In a 2019 article for the publisher Elsevier, Alex says they are always "trying to lead with my most authentic self, and whether you take that or leave that is totally up to you."

When Alex was a kid they watched a ton of Discovery Channel and National Geographic and were always asking questions about everything. "I'm sure my family found that very annoying!" they say, laughing a little. "I wouldn't say I realized I was being a scientist." But it's true: this is what scientists do. They ask a lot of questions, and keep looking for answers.

Alex enjoyed science and took lots of science classes in high school. They say, "I just knew that I liked these classes, not that I wanted to do this for the rest of my life." In college, they switched majors a few times at first, from chemistry to biology, then finally to **ecology** and evolutionary biology. For Alex, finding the right area of study was like Goldilocks . . . finding a research area that's not too big, not too small, but just right. The giant scale of evolution was too large for them, but at the same time, some of their graduate lab work focused on just one species, which felt too limiting. Looking at interconnected

WHAT IS ECOLOGY?

Ecology is a broad area of study that includes lots of fascinating subjects. At its simplest, it is the study of the relationships between living organisms, including animals, plants, humans, and the environment where they exist. Some ecologists study plants and animals in deserts or arctic settings, while others might study invisible parasites in a particular species of bird. There are many different ways to specialize in ecology, and for many people, like Alex, it might take some time to decide exactly what to study. But ecology plays a really important role in all parts of our world, from helping manage wildfires to protecting endangered species, from looking at how we treat disease to looking at how we treat the planet.

"But for me the most difficult part was finding community and finding like-minded people who have shared experiences."

ecological areas, studying how different species intersect and impact each other, provides Alex with that "just right" fit.

"I became more and more curious, and more and more excited about the things I was learning, and that kept pushing me forward."

Even as Alex discovered what they most wanted to study, pursuing their career in science wasn't always easy. They say, "There are always academic challenges. The work is hard. But for me the most difficult part was finding community and finding like-minded people who have shared experiences. Being the only person of color in the department is super challenging, all on its own. So finding space where you feel comfortable and finding support networks can be tricky because it's not really built into the system at most schools. And there's no real guidance on how to do this." Alex believes that finding or building a support network is one of the most critical things a student can consider before choosing a school.

Alex acknowledges that while more and more colleges are trying to recruit diverse students, they are not always making the effort to ensure that those students feel welcome and comfortable. Alex talks a lot about the need for **code-switching** in these spaces. In the article for Elsevier, Alex says, "When I walk into a room, all you see is that I am Black. So that is something I'm very aware of. And most of my life has kind of been about code-switching. . . . But it's been really important for me to also push forward the rest of my identities." For Alex and many other folks who have several identities, such as Black and queer, there are constant decisions about how to present themselves to new people, and

WHAT IS CODE-SWITCHING?

Code-switching refers to a person changing how they speak, act, dress, or behave depending on whom they are interacting with. More specifically, the term can also refer to the way that Black people modify their behavior to avoid discrimination or stereotyping by white people in their schools, communities, and jobs. This is important because code-switching continues to reinforce the idea that typically white habits of conversation, dress, or behavior are the "normal" or "good" ones, instead of acknowledging that these habits are simply the most comfortable for white people.

When Alex talks about code-switching, they are acknowledging something that many Black people do at work or at school, while shopping or eating in a restaurant . . . really anywhere that requires interacting with a white community. Studies have shown that code-switching helps Black people be seen as more professional and can help them in their careers. But the truth is that hiding part of your personality and changing your speech, behavior, and even your hair to help other people feel more comfortable is exhausting and takes its toll. As Alex says, true equity will happen on college campuses not only when there are more students of color, but when all students are able to be their authentic selves and still be taken seriously as scholars and learners.

how much they want to share, especially when it comes to identities that can be less obvious at first glance like gender or sexuality.

At the same time, Alex is sure they made the right choice. "The best part of my job is that I'm doing what I want to be doing. I'm motivated to do the work and I get to do what I set out to do." When talking about their job, Alex is clear that there's a wide variety. They conduct research at field sites from New England to the South Pacific, teach high school students through graduate students, and present at conferences. They say, "In fieldwork season, I am planning my travels out to the field, then going and collecting samples, analyzing those samples in the lab, and writing up the results."

The actual science Alex uses includes a range of techniques. Some of them are cutting-edge, involving lab equipment that can test how much nitrogen or phosphorus is in a soil sample. Testing for nutrient content in the soil can help scientists understand what nutrients are available for plants to use for

When asked how they would describe their job to their younger self, Alex says, "I would tell myself that I spend my time teaching people why we should care about the environment, why we should care about specific habitats and the ways we humans impact them. And I'd also say that in my research I'm interested in understanding how different types of animals interact with each other, and hurt or help each other, and how they share spaces.

"I also study the ways humans fit into the puzzle. But not always in the ways people think. One major thing I look at is how different cultures interact with the environment. How have people used these ecosystems in ways that have historically been helpful and sustainable? How can we learn from them?"

their growth, or whether some environments are lacking important nutrients or have a surplus, which can cause damage.

Other techniques Alex uses are so old that cave dwellers could have used them. They explain, "One of the most primitive techniques involves trying to understand how much organic matter (which is made up of the remains of plants and animals) is in a soil sample, versus how much is mineral, or rock. The easiest way to do this is to burn it! You could literally use a campfire, but we use a high-heating oven in the lab." Since organic matter burns, and rock or mineral does not, Alex measures the difference between the sample's weight before and after burning, and can tell how much of it was organic. This tells scientists information like how much carbon is stored in the soil, which can be really important for addressing climate change, or how much organic matter is available to decompose into nutrients that support plant growth.

But Alex is clear: whether high-tech in the lab or lighting things on fire, they use the same basic techniques that they learned in school. It all starts with the scientific method—asking a question, then coming up with a hypothesis and testing it.

In contrast, when it's not time for fieldwork, Alex says they spend a lot of time teaching in the classroom, but also coming up with new research ideas, writing proposals to get funding for the ideas, and doing data analysis in the lab. Then when fieldwork season arrives it starts all over again!

For Alex, who is nonbinary and is part of the LGBTQIA+ community, as well as being Black, giving advice to others who might be interested in science is complicated. "I don't want to say 'Just do it!' because there are some spaces that could be really terrible. But at the same time, if you have an interest in science you should definitely investigate it! Don't be afraid to explore just because you don't know anyone who has done it before. And find people who will support you through the process. You can't assume that you'll go into a space and be made welcome, but you can find support. I think my advice would be to be more intentional about finding safe places and a community of people."

ADVICE FOR YOUNG PEOPLE
"As you think about colleges, don't just look at the subjects they offer. Look deeper. What does the faculty look like? What does the student population look like? Are there resources on campus—like cultural centers, affinity groups, and support groups? Are there resources for people who might be unprepared for college, like first-generation students? These are all things that will help you understand if you might find a community there."

Alex also thinks that it is never too early to talk to any college students you know about their experience. They add that it's also useful to talk to students even if they're not studying exactly what interests you. They can talk about the culture on campus, and how easy or hard it is to find community. And Alex also urges students, when they get to college, to connect with folks beyond their own department. "There are so many people passionate about different kinds of science, and it's worth finding people who understand your shared experience outside of your silo.

"Because finding that community was the hardest part for me, I wish someone had told me that I needed to be intentional and seek it out. My experience would have been very different if I had made that a priority. Instead I isolated myself and just tried to get my work done, and that was hard."

Despite the challenges, Alex loves their job. One of the best parts, that they could never have imagined, is how much traveling they get to do. They say, "I travel to places I never could have dreamed of. The way I grew up, I never thought about travel like this, and never realized that an option like this existed."

Alex is currently doing research in American Samoa, a Pacific island between Hawaii and New Zealand. They have also spent a lot of time working in Hawaii, and have gone to Europe for conferences in England, Scotland, Belgium, and France, as well as to Brazil and Argentina in South America! As they put it, "This kind of travel is life-altering."

> ## "I spend a lot of time collecting mud from random places."

One of the weirdest parts? Alex's job involves something many of us did in our backyards or local parks: they play with mud! Alex says that wherever they travel, whatever the field site, "I spend a lot of time collecting mud from random places."

Alex again reiterates that while science involves facts and data, that doesn't mean there aren't biases, stereotypes, and assumptions made. Society's values and the political culture can change what gets prioritized in science. They say that until recently "there has been a shift away from valuing ecosystems, and that is really frustrating. We see this in the removal of species from the Endangered Species Act, and the proposal to drill oil in the Arctic." While the proposal to drill in the Arctic National Wildlife Refuge has been suspended for now, Alex points out that the fact that such a fragile ecosystem was considered for drilling is "the opposite direction from the way we should be moving."

Alex also recognizes that some things are getting better. They notice that while historically mainstream science has not valued or paid attention to traditional ecological practices, or the ways that Indigenous people managed their ecosystems, that is changing. For example, Alex says, "Typical Western science says the only way to protect a landscape is to keep people out. But there are communities that have been able to live with the land respectfully and successfully, and who have connections to that land for both their culture and their livelihood. So there's more to learn."

The challenges of climate change show up in Alex's work all the time, and they know there are no easy answers. Much of their work is studying the coastline, where land and sea meet. They point out that almost 40 percent of the human population lives on the coastlines of the planet! And with increased flooding and sea-level rise due to climate change, there is a lot of research on how to keep these spaces safe.

Alex says, "We are looking at understanding how we can use **mangroves** or **salt marshes** to help buffer storms, and how we can manage these landscapes in ways that help filter pollutants. But the challenge is that we have to conserve and restore these habitats in a way that they can adapt to what's coming, and what's changing. It's not that we're predicting the future, exactly, but something close to it."

"Trash is one of the largest detrimental impacts that the average person has on ecosystems and spaces. We need to think about what we buy and throw away."

"If people commit to something, if we work together, we can actually make things better."

Similar to mangroves, salt marshes play a vital role in nurturing plants and animals and protecting coastal areas from flooding and sea-level rise. Alex and their colleagues did a study of predators, like green crabs, in the salt marsh. They marked off areas with mesh cages to analyze whether the absence or presence of these predators influenced the ecosystem. They looked at soil organic matter content, soil nitrogen content, plant biomass, and more to understand how different species influence the wetlands.

Trash is an inevitable part of life, but we can all be more aware and careful with what we buy, and how we dispose of it.

One of the major actions Alex hopes we can all take to help the planet is to be more mindful about how we consume things and get rid of waste. "Trash is one of the largest detrimental impacts that the average person has on ecosystems and spaces. We need to think about what we buy and throw away." Trash is an inevitable part of life, but we can all be more aware and careful with what we buy, and how we dispose of it. Literally tons of garbage wind up in the ocean, and the currents swirl it together, creating horrific floating garbage patches. The largest of these is twice the size of the state of Texas! All of us can influence our families, schools, and communities to help improve how we all deal with trash. The first step to take, and the most useful, is to limit the amount of garbage we're creating. We can all look for alternatives to single-use plastics and try to buy reusable or compostable items instead. If disposable plastic is required, we can help recycle as much as possible. And finally, we can all ensure that garbage gets thrown away properly, so that it doesn't blow into the water and wind up out to sea. As Alex says, "If people commit to something, if we work together, we can actually make things better."

Alex also says, "I want to remind people that it's not all doom and gloom. I am excited about what's happening right now. I think there's an interest in Indigenous learning and land use that complements Western science, and I'm excited to see people realize how these can work together, and move us toward the goals we're looking for in a way that is more inclusive and equitable to all."

WHY ARE MANGROVES AND SALT MARSHES IMPORTANT?

Imagine a technology that can filter pollutants out of seawater to make it cleaner. Then imagine this tech can also protect dozens of endangered plant and animal species. Finally, what if this incredible invention could also help keep soil from eroding and protect the shoreline from sea-level rise and storm surge? Amazing, right? Well, nobody needed to invent this mind-blowingly useful technology, because it already exists, growing happily in environments around the globe.

Mangrove trees are found all over the world, in areas where land and sea meet. There are over eighty different types, and they all grow in low-oxygen soil, where not many other trees can live. Mangroves are sometimes called walking trees, because it can look like their roots are walking on the water. But these aboveground roots, also sometimes called breathing roots, work to provide oxygen to the tree despite poor soil.

Similarly, salt marshes also exist around the globe. They provide habitats for vital ecosystem populations that support the whole aquatic food web, including birds, fish, mammals, and more. But they also help humans more directly by protecting inhabited areas from storm surges and rising sea levels. Both mangroves and salt marshes are some of our best tools to protect coastlines and endangered species, and scientists like Alex study how they work and how we can best protect them and help them do their job.

"I want to remind people that it's not all doom and gloom."

"I blaze my own trail."
—Anita Marshall

ANITA MARSHALL

Anita Marshall keeps surprising everybody, including herself. Though she loved *National Geographic* magazine, she hated high school and had to be "dragged kicking and screaming" to college. When her high school teacher got permission to teach a one-semester geology class, she approached Anita, who asked her what geology even was. The teacher answered, "Anita, it's everything you love! It's volcanoes and earthquakes and landslides and rocks!" Anita did love the class, and credits that teacher for making it interesting. But she also notes that the teacher never once made the connection that this was actually a career people had. Geology was taught like a fun hobby—something that folks could do on the weekends by going out rock-hunting.

◀ Anita at the Sunset Crater Volcano National Park in Arizona, where there are several wheelchair-accessible trails that wind through the lava flows at the base of the volcano. The mountain behind her was formed around 1,000 years ago, during the most recent volcanic eruption in Arizona.

Anita (left) and her sister and cousins at the Garden of the Gods in Colorado. This photo was taken when Anita was in the tenth grade, around the time she discovered her love of geology.

As the daughter of a military chaplain, Anita changed schools fourteen times before she graduated from high school—a fact that she realizes played a role in her dislike of school. She went to college to study music education, planning to be a band director. As a member of the Choctaw Nation, Anita and her family had a complicated relationship with education. Anita's great-granny had been enrolled in an Indian residential school, where they forbade her from speaking Choctaw or practicing her tribal traditions, and Anita's grandmother dropped out after eighth grade, saying, "High school doesn't put food on the table." Her father attended a Bible College on the GI Bill, the first in the family to go to college.

Anita chose the University of Arkansas because it offered tuition discounts to all the Oklahoma tribes along the Arkansas border. She had no idea that it was a school with a great geoscience department. Still, when she had to take one science class, she remembered that high school elective and found a geology class. Then, she says, "Two things happened. First, I had to choose between sitting in a windowless practice room practicing my instrument, or going on a field trip. And that was easy! But second, I met people who were geology majors. And for the first time I realized I could get a degree in this. I could do this for a career! It was the first time I understood that I could be a scientist."

She laughs. "I should also say that my first college geology professor was the worst! If I had not taken that high school class I would not have been hooked. But it was meeting my peers, meeting students who were studying this stuff and making a career out of it . . . that's what got me interested."

> "In order to get a degree in geology you need to be able and willing to disappear into the wilderness for six weeks."

One of the greatest things about geology, for Anita, was the opportunity to be outdoors in the field. "In order to get a degree in geology you need to be able and willing to disappear into the wilderness for six weeks," Anita said. "And doing this kind of fieldwork is my favorite part of geology. I love getting out and learning more about how the Earth works."

But still, Anita wasn't sure she wanted to follow the traditional path. As college wound down, all of her fellow geology students were applying to graduate school, and she wasn't sure it was for her. In fact, she had a professor tell her straight out that she wasn't smart enough for graduate school. She says, "He shook my confidence. It's hard when these authority figures tell you, 'You don't have what it takes.'" Anita wasn't willing to apply to the prestigious schools her friends were applying to, but she did apply at her local university.

Anita worked on her master's degree, traveling to the Caribbean to do fieldwork. But then, right before her final semester of school, everything changed. Anita was in an accident, hit by a drunk driver while she was walking. She spent two months in the hospital, then had many months of reconstructive surgeries. She used a wheelchair for a year, had many more years of physical therapy, and still has significant disabilities from the accident.

She says, "In hindsight, my commitment to finishing my degree on the same timeline was probably not a great decision. I was going into surgeries on a regular basis, and when the spring semester started I was still in a rehab

WHAT IS GEOLOGY, AND HOW DOES IT RELATE TO CLIMATE CHANGE?
Geology is a really broad area of scientific study. Not only does it include pretty much everything that makes up planet Earth, from rocks to soil to water to ice, it also studies all of those elements from the beginning of time, going back billions of years into the past. Geologists often specialize in a certain area, whether that's glaciers, the ocean floor, volcanoes, or earthquakes. But since all elements of the Earth are interconnected, a study of geology often involves following lots of different threads and learning lots of different things.

Studying geology means having a front-row seat to the understanding of climate change. Since geologists study the way elements of the planet change over time, they are aware that Earth's climate always changes, and often with dramatic temperature shifts. Geologists find clues about the changing climate everywhere, from the rings of old trees to the microscopic fossils found in ice samples drilled from deep in Arctic glaciers. The story of our planet's temperature can be found in the rocks and sediment and ice all around us.

But while climate is always changing, it has become clear that the way humans live on Earth right now is speeding up climate change in new and dangerous ways. In order to grow the crops that feed us, and live in communities that are not too cold or too hot or too dry or too flooded, we need to keep the Earth in balance. Geologists of the future will read the clues in the earth, just as geologists do now, but what kind of damage they find will depend on how we take care of the planet today.

"I was committed to finishing my degree, but I really thought I was done in geology."

hospital, but I really wanted to do my classes." While she started the semester a few weeks late, Anita was determined to get back to school.

Anita's whole family supported the effort. Her dad had quit his job to help care for her, and together they managed to rig her grandfather's power wheelchair together with a kind of traction sling for her leg. Anita says, "I was committed to finishing my degree, but I really thought I was done in geology. After all, I wanted to be a field geologist and there was no way I was going to be able to do that. So this was my swan song. . . . It was a way to close my very short run in geology."

She continues, "I had no plans beyond finishing that degree. I just didn't want to be a quitter. But as I was wrapping up, one of my favorite professors asked if I would be interested in teaching. I had never taught before, but the local community college needed a part-time professor, so I said yes."

Anita taught for several years and was able to keep working on her recovery while staying involved in geology, passing on her excitement to her students. Each semester, she says, she showed the same video to her students, a movie called *Volcano: Nature's Inferno*. In it there is a short clip where a world-famous volcanologist who had been injured on the job returns to the site of an active volcano. Anita says that after watching the movie dozens of times, she focused in on that one minute of film. "I thought, he's got injuries a lot like mine, and he's still out there studying volcanoes. Isn't *that* interesting!"

Of course, the famous volcanologist in the movie had gotten his injury after building his career and becoming famous. So that became Anita's

WHAT IS ABLEISM?

Anita experienced **ableism** after her accident as she learned to work within the new restrictions of her disability. This term refers to discrimination or prejudice that disabled folks experience. It is based on the belief that typically-abled people are more worthy, and any accommodation for disability is an inconvenience or sacrifice. Ableism can show up in lots of different ways and can relate to both physical and mental disabilities.

question: "Can I do this if I haven't already earned my stripes?"

She started looking at graduate programs where she could pursue her PhD in geology, and found that unfortunately her concerns were well-founded. "There was a lot of **ableism** and a lot of pushback," she says. "There's a sense that if you can't haul fifty pounds of equipment on your back up a mountain, you're not a geologist."

She eventually got into a program at the University of South Florida, and entered an era of her life she calls "the Volcano Ventures." Specifically, she was part of a geophysical volcanology group that used a variety of tools to study all kinds of volcanic features. Those included ground-penetrating radar, gravity measurements, magnetic measurements, and other things to understand what's underneath the surface of a volcano.

"I absolutely loved this aspect of geology," she says, "being able to almost magically see what's beneath the ground. But there were colossal barriers to my being able to participate the way my friends and colleagues were. There were enormous social barriers, and there was a stigma attached to disability that was the elephant in the room with almost every project I participated in."

It's not just physical disability, Anita says. "Geology is a field that prides itself on being outdoorsy and rugged and adventurous. For people who don't fit that description for any reason—because they grew up in cities and never camped a day in their lives, or because of physical ability—there's a lot of pushback."

Still, she didn't want to give up. "I was pushing myself beyond safe limits, putting myself in dangerous situations, just trying to keep up and do things the way everyone else was doing it."

Then on one field trip Anita ended up getting left on the side of a volcano in a jungle in Nicaragua, and she realized: there had to be a better way to do this. She started looking for other disabled people to learn how they were handling the challenges.

"It was hard to recognize that the part that I loved the most—fieldwork—was the single biggest barrier to degree completion for people with disabilities. The fact that they couldn't do field camp was one of the main reasons students with disabilities were dropping out, or what we call 'self-selecting' out of geosciences." She goes on, explaining that fieldwork is important not only because it allows students to take complex classroom ideas out of their textbooks and apply that knowledge in the real world. In addition, she says, "Fieldwork is how students start to feel like geologists. It gives them confidence and a sense of identity."

At this point in her studies, Anita started pouring a lot of her time and energy not just into studying volcanoes, but also into studying accessibility in the geosciences; in effect, studying how more students of various abilities could do this work. She began asking questions: "What's been done to improve accessibility? What *hasn't* been done? Who is working on this issue?"

Before long, Anita found herself essentially pursuing two PhDs at once: one in volcanology, and one in geoscience education and accessibility. Not everyone was

"What's been done to improve accessibility? What hasn't been done? Who is working on this issue?"

encouraging. "I had a professor tell me to drop out," she admits. "He thought I was wasting my time on accessibility issues. I did not get a lot of support. Unfortunately that wasn't surprising. All of geology has an ableism problem, but volcanology is at the extreme end of that."

As she was trying to integrate her two areas of study into one PhD project, a friend shared information about a program that was looking for undergraduate students to help with a large-scale project on accessibility and science. She laughs. "This was either the bravest thing I've ever done or the most ridiculous thing, because I had never met these people, and I wasn't really what they were looking for, but I sent them an email out of the blue, saying I'm a graduate student, not an undergrad, but I wanted to be a researcher on their project." They responded quickly, saying, "It sounds like you were born to work on this." And, as Anita says, that was the start of a long and beautiful friendship with researchers she still works with to this day.

After completing her PhD, Anita took a job at the University of Florida, a school that she says was excited about accessibility work and supports all aspects of her research. She spends the majority of her time working on accessibility issues. Volcano research takes up only around 20 percent of her time. But, she says, "That's okay. Because at the end of the day, while my work on volcanoes is important, my bigger impact is in helping make the field more inclusive." Because so few people are working on accessibility for people with disabilities, Anita gets a lot of calls and emails. She says, "If I can help disabled students find a place in geology, that's what I want to do. When teens reach out hoping for answers, I'm going to help."

She points out that "for people looking in from the outside, it seems like I had a nice smooth trajectory, but it was rough getting to this point. I had to blaze my own path. So I want to help others and make it easier."

For Anita, connecting with other disabled geoscientists, even virtually, made all the difference. "It was freeing to be with people who understood

ADVICE FOR YOUNG PEOPLE

Anita says, "Probably my biggest piece of advice I'd tell students with disabilities is don't assume you can't do that thing, just because people do it a certain way. That's a very narrow lens, and very limiting. I really started thriving when I realized there's no reason to do things the way everybody else does them. I can figure out a different way. I do my fieldwork differently than my able-bodied colleagues, but I get the job done."

She continues, "I strongly recommend prioritizing good mentors over a specific area of study. No matter how interested you are in an area of study or career path, don't go anywhere if the people aren't supportive. Ultimately people will be far more important and influential to your development as a person and as a scientist. A mentor doesn't need to be someone who matches your identity, so much as it needs to be someone who is open and supportive and flexible."

On the other hand, she says, "When it comes to your peer group, it's really hard to be the only one. The question of identity becomes more important. You need people around who understand your experiences, whether that relates to your disability, race, or any other part of your identity. Otherwise you start to feel like you're all by yourself."

the challenges, and who can help each other." Even while being at different schools all around the country they are able to offer support and community, and that makes all the difference.

Today when Anita gets to study volcanoes, she doesn't go into the danger zone of an eruption. Realistically, even able-bodied people can't outrun an erupting volcano, she realizes, but she says that psychologically, "It bothers me to know that a brisk walk is the best I can do if disaster strikes!" But she doesn't feel like she's missing out. "Studying quiet volcanoes is just as important, because we can get all the details, and take as long as we want. And when we get a really good understanding of how a quiet volcano works, we can apply that knowledge to other erupting ones."

When talking about the scientific tools she uses, she says, "At its heart geology is an observational science. We are keen observers, paying attention to detail. And for the first hundred years of geology, it really was just people observing natural processes and taking detailed notes. Which is amazing,

"Volcanologists have always looked for ways to increase accessibility. . . . When the conversation shifts to including disabled people, there's suddenly a line drawn about what degree of accessibility we're willing to consider. And that's where I'm pushing back."

In Connemara, Ireland, Anita (right) and colleague Dr. Jen Piatek (left) made gigapan images: high resolution images used to create virtual field trips or study tiny details in rock. They are using big zoom lenses on their cameras because neither of them could traverse across the bog to get closer to the rocks.

because anyone can be a geologist! We can all be observers who pay attention, then try to figure out what happened and why. And while we've added some cool tools, like radar devices, ultimately it's still observational."

One of the interesting things she has realized in her combined work, studying volcanoes and working for more accessibility for disabled students, is that volcanoes have always been inaccessible. "Working on active volcanoes is incredibly dangerous. People die doing this work. And volcanologists have always looked for ways to increase accessibility to places we can't get to, and to make it safer."

As an example, Anita explains how scientists can now send drones to capture gas samples above an active volcano. It used to be a risky process that involved challenging hikes, four-wheel-drive vehicles, or sometimes even sending a helicopter above an active volcano and having a person literally climb out of the copter and try to grab a sample. Now with drones, scientists can work around a barrier to access and get more and better information.

But, Anita says, "When the conversation shifts to including disabled people, there's suddenly a line drawn about what degree of accessibility we're willing to consider. And that's where I'm pushing back. I'm reminding the geology community that they were already thinking about ways to overcome access barriers, and now we can apply that same mindset to including disabled scientists." She points out that while geologists like to think of themselves as Indiana Jones characters, new technology is changing how geology works, and there are new fields of study where people are using computer models and coding in a lab instead of scaling mountains or running around active volcanoes.

The most important thing she wants younger students to know is that yes, you can make a career out of this! "I didn't know until I got to college that this was something I could study," she says. "Algebra and calculus are the math languages of the geosciences, so learning algebra, then on through calculus in high school, is something you'll keep using." She continues, "I'd also remind kids that while volcanoes might seem far off, there are natural hazards in all our environments, whether they're tornadoes or hurricanes or wildfires. And climate change is the big force multiplier. It takes all these hazards that already exist and pours gasoline on the fire. So we have to address that things are changing, and what happened historically might be different from what's going to happen." She notes that when she teaches college students, some are very concerned about climate change, and others don't have a good understanding of the problem. "You never make friends by telling people they're wrong and what they believe in is wrong," she says, so she works to help all students understand how Earth is reacting to the changing climate, giving them a better understanding of what's happening on our planet. Ultimately she wants to bring as many people as possible into science.

"Collaboration is a powerful means for inclusion, and leads to stronger research," Anita says. "To succeed, everyone needs to feel welcome. Every part of you can make your science stronger. Everything from my family's history to my challenges with recovery and physical disability informs how I do my science. I think it makes me a stronger scientist. So don't be afraid to embrace your whole self in science . . . bring all of you into your work, and if you can't see an obvious fit, blaze your own trail!"

"Don't be afraid to embrace your whole self in science . . . bring all of you into your work, and if you can't see an obvious fit, blaze your own trail!"

"I think of myself as someone who likes mysteries, who likes unpacking things and discovering things, and asking questions." —Rae Spriggs

RAE SPRIGGS

Rae Spriggs has met with presidential candidates to talk about climate change, graduated from a world-renowned university with a degree in environmental health science, and worked on dozens, if not hundreds, of proposals for new science projects. But she still hesitates to call herself a scientist.

"I don't know what it is about the title of 'scientist,' but I struggle to identify myself that way."

Instead, Rae says, "I think of myself as someone who likes mysteries, who likes unpacking things and discovering things, and asking questions. Science is a lens to unpack all these mysteries and questions I had."

Rae grew up fascinated with science, especially the planets and the solar system. And she was lucky enough to grow up with a lot of support and encouragement. Her pediatrician, from the time she was a baby, was a Black

◄ Rae thought she wanted to be a doctor, but when she learned about environmental justice and how our health is impacted by the environment, she was hooked.

woman, and Rae says, "I looked up to her. Even then, representation was so important to me. And I was really fortunate to have amazing teachers. I probably owe a lot of my success to them." Rae grew up in San Diego and went to an elementary and middle school that was led by a Black principal and had mostly Black teachers. She says, "They really affirmed me. They never let me say negative things about myself—they just didn't tolerate it."

Rae grew up wanting to be a doctor, and when it came time for college she got scholarships to attend the prestigious University of California, Berkeley. But college life was a profound challenge. Rae went from being surrounded by other Black and brown faces to often being the only Black person in her classes. And her first chemistry class?

"I didn't get the grade I needed to for premed. I dropped the class," Rae says. "And just like that, I went from thinking I loved chemistry and medicine, and thinking it's my life path, to thinking, I don't know if I can do this."

Much of the hardship, she says, was that the environment on campus felt unwelcoming. While many students from all different backgrounds fail the notoriously difficult first-year premed classes, Rae says that for students of color, the feeling of not belonging can make those early setbacks much harder. After wanting to be a pediatrician for much of her life, Rae started to question whether this was a path she could follow. But during her sophomore year of college, she took an elective: a bioethics class that focused on issues of **environmental justice**. And Rae was hooked.

"I had never heard that term, 'environmental justice,' before. It was the first time I thought about the connection between our health and the environment."

WHAT IS ENVIRONMENTAL JUSTICE?

Many of us know what **social justice** means. It's a way of talking about efforts to improve society's systemic problems like racism or poverty. And **environmentalism** is pretty clear: it's a movement to help improve the environment and the planet, by reducing pollution and garbage and protecting the natural world.

But let's talk about **environmental justice**. It is the meeting of these two important ideas, where social justice and the environment are linked together. Environmental justice deals with problems caused by human activity, which are only getting worse as our climate keeps changing, and how these problems are more serious for some populations than others. While the challenges of climate change will eventually affect everyone on the planet, that doesn't mean that we are all experiencing it in the same way. Environmental problems such as air pollution and contaminated water are worst in communities that are oppressed by racism and poverty. And more recently, the parts of the world that are most at risk from extreme weather like flooding are in countries that have done the least to contribute to climate change! And this has not happened by accident.

For years and years, in the United States and around the world, the most polluted areas have been places where poor people and people of the global majority live and work. Corporations and governments have chosen these communities to build landfills, highways, factories, and toxic-waste dumps, and the folks who lived nearby have suffered the health consequences. Both the land and the local people are harmed. And now as the effects of our changing climate are getting worse, the most vulnerable communities are the ones that will get hit the hardest. So while we might all be in the same boat when it comes to climate change, some of us are on the top deck eating chocolates, and others of us are barely clinging to the sides, already slipping into the water. Nobody would call that justice.

Environmental justice is a movement to address the challenges of our planet while also recognizing that all people, not just those in power, matter. It means that everyone should have access to clean water and air, and that everyone deserves a healthy environment to work, live, and play. It means that as we search for ways to help meet the challenges of climate change we must consider the needs of all communities, not just the wealthiest. More and more, social activists and environmental activists are banding together, recognizing that the planet and the people will do best if we all work together.

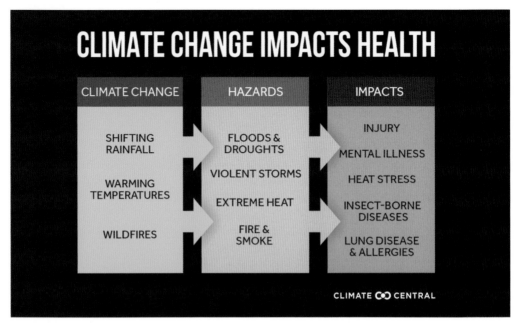

CLIMATE CHANGE IMPACTS HEALTH

CLIMATE CHANGE	HAZARDS	IMPACTS
SHIFTING RAINFALL	FLOODS & DROUGHTS	INJURY
		MENTAL ILLNESS
	VIOLENT STORMS	HEAT STRESS
WARMING TEMPERATURES	EXTREME HEAT	
		INSECT-BORNE DISEASES
WILDFIRES	FIRE & SMOKE	LUNG DISEASE & ALLERGIES

CLIMATE ∞ CENTRAL

The impacts of climate change on health are serious. This chart shows the effects a changing climate can have on people.

"I had never heard that term, 'environmental justice,' before. It was the first time I thought about the connection between our health and the environment, specifically the air we breathe, the water we drink, and so on. It was the first time science came alive for me."

The class felt personal for Rae. As she dug into this concept of environmental health she recognized that she and her brother, who grew up right next to a giant highway, could trace their asthma and respiratory problems directly to the pollution from the cars that drove by. She says, "Something really sparked in me then. It made me angry. I grew up breathing this **particulate matter** and having health problems associated with it, and my story was not uncommon. I started to question why we as a society are allowing this to happen. Why didn't we know about this? Why were our systems keeping us, my family, my community—the people on the ground—from knowing the potential health impacts of our environment? I wanted to fix it."

Newly motivated by her interest in public health and environmental justice, Rae powered through her four-year degree at Berkeley in three years. Even though she was excited about what she was studying, Rae says she never felt at home there and was eager to graduate. She points out that the years she was there were tumultuous ones, when the killings of Black teens like Trayvon Martin and Michael Brown were making national headlines and leading to protests and counterprotests. She says, "Campus was a hostile place. There were protests right outside the dorms I lived in, and sometimes the tear gas used by the police was so strong it seeped into our dorm rooms. There were hate crimes on campus and a sense of unrest, and I really wanted to get away."

WHAT IS PARTICULATE MATTER?

Particulate matter refers to tiny liquid droplets or solid particles that are found in our air. They are small enough to be inhaled into the lungs, and sometimes even small enough to get into the bloodstream! Common sources of particulate matter include direct sources, such as smoke or construction sites, and indirect sources, like exhaust from cars that burn gasoline or pollution from power plants. Inhaling particulate matter can make people really sick, and, as Rae learned, unjust environmental regulations have often led to the worst pollution happening in the poorest neighborhoods.

Before graduate school, she played a leading role in researching state and national best practices for assessing climate change and health equity. In this job she drafted policy on health and the community for the county of San Diego. The information she gathered was used to understand the quality of life in different neighborhoods. These projects confirmed for Rae that the work she loves best is at the intersection of environmental science, public health, and political policy.

When Rae arrived at the University of California, Los Angeles (UCLA) for her master's program she found a campus community of support that she had been lacking at Berkeley. Reflecting back, she says that one thing she wishes she had realized—as obvious as it sounds—is that when you are considering colleges, you are going to actually be living there. She says, "I would encourage high school kids, and especially POC [people of color] kids, to look into what it's like to live on campus. Are there affinity groups of diverse students? Are there spaces for people to gather? Make sure there are support systems in place."

At UCLA, Rae found a community of other students of the global majority studying all different fields in public health. Together they organized public health events and activities, building a support network as they pursued their degrees. During these years, Rae also found an incredible mentor in her adviser, whom she continues to work with. Interestingly, she says that her first adviser, who was the only Black woman in the department, was ultimately not a good professional fit for Rae. She notes, "It's important to have representation, but it's also important to ask questions about whether you're aligned on the science, and who their stakeholders are. For me, the biggest stakeholder is the community." Rae's first adviser was more interested in pure scientific inquiry, and Rae realized that she would be better matched with someone who focused on the intersection of science and community health. Rae switched advisers, to a white male professor whose work was more aligned with her interests, and who she says to this day remains a great mentor and part of her support system.

Again and again in her work, Rae saw the connections between problems in public health and political policies that allowed projects with negative environmental impacts to be located near communities of color and low-income communities. Pollution from factories, cars, and landfills, among many other sources, caused diseases from asthma to cancer at higher rates in these communities. Rae's own health issues, and those of her family members, were squarely at the center of a larger problem. She realized that instead of studying to be a doctor and helping a handful of patients, she wanted to maximize her impact by working on science and policy that could help whole communities.

During graduate school, Rae worked as a climate and equity researcher at the Nonprofit Institute of the University of San Diego. There she built a set of data points based on regional indicators that related to climate change and equity in communities. The information she gathered was used in the San Diego Regional Quality of Life Dashboard, a tool that helps San Diego regional stakeholders (residents, businesses, nonprofits, government agencies) understand changes in the quality of life in San Diego over time. They study areas ranging from air quality, beach and coastal health, and employment and income, to climate change and planning. While at UCLA, she also researched health impact assessments, or HIAs, which are tools that decision makers like politicians, public health officials, and developers can use to figure out the potential impact of different projects. The goal of the HIA is to find ways to reduce harm, such as pollution from a highway, and maximize potential good, such as building or protecting a park in an underfunded community.

"We are done accepting business as usual—it's time to awaken minds and change the systems."

Rae at graduation. She received her master's degree in public health with a concentration in environmental science.

"I have always been drawn to jobs where I'm the first one, where I'm a trailblazer who has to build the role while I'm working."

Rae's job now is associate director at the Center for Diverse Leadership in Science (CDLS) at UCLA. In this role she wears a lot of hats, and juggles a lot of different elements. Instead of working on one research project, Rae develops community outreach programs, builds partnerships with local community health organizations, and supports and helps develop lots of different research projects that graduate and postgraduate students, known as early-career fellows at CDLS, are working on. These projects might be in environmental or ocean science, or civil engineering. One example of a project Rae is helping manage is by a CDLS fellow who is studying the health impacts of the brownfield—whose soil is contaminated with toxic waste—that is located right near a Los Angeles housing project.

IF I HAD A TIME MACHINE
"If I were describing my work to my younger self, I would say that I am in a field where we study environmental components like the air we breathe, the water we drink, the land we build on, the parks and green spaces we play in, and the changing climate."

Rae says that for her, an average day can be writing grant proposals to get funding for new projects, or reviewing existing reports on issues of environmental justice, or meeting with groups of young scientists to work on a collaborative project, or meeting with community organizers or potential funders. It's a lot of juggling, but Rae loves it.

"I have always been drawn to jobs where I'm the first one, where I'm a trailblazer who has to build the role while I'm working. I like to be the person who figures out what's possible."

Rae says that the issues of public health and environmental justice have definitely grown more urgent with the impacts of climate change. She says when she first got involved in this work, she thought about environmental health in terms of pollutants and cancer-causing agents in the air, water, and soil. But, she says, "Now it is clear: climate change is a huge driver of environmental injustice and worsening health outcomes for lower-income communities and communities of color. These are the areas that often have the least resources to adapt to our changing climate, and are the most vulnerable to threats." She adds, "We are done accepting business as usual—it's time to awaken minds and change the systems."

One new development that Rae is excited about isn't a scientific breakthrough, but does have the potential to dramatically shift the focus and impact of environmental science. Rae is part of a movement to get more scientists of color involved in environmental science and geosciences, an area that has very low levels of diverse participation. "The more momentum we get, the more of us there are, the better the science will be." She points to how few faculty of color there are at most universities, and how that can make students of the global majority feel unwelcome. "Environmental and geoscientists from schools across the nation are advocating to diversify our fields. We are organizing calls to action on campuses around the country, and hosting anti-racism workshops so that university faculty can better support Black and POC scientists."

"The more momentum we get, the more of us there are, the better the science will be."

"Kids have magic—they can ask questions and make something important just by asking."
—Rocío Paola Caballero-Gill

ROCÍO PAOLA CABALLERO-GILL

Rocío Paola Caballero-Gill has pivoted many times in her career, moving across continents, changing career paths, and contending with physical illness and setbacks. But through it all she has remained committed to the same belief: that life is best spent continuously learning new things, then using them to give back to your community.

As a child Rocío was always curious, and always loved playing outdoors or in the dirt. She says, "Even when I had my Barbie dolls in Perú, they were blond and looked nothing like me, but I always chose the one dressed like an astronaut, because I liked the tools and the outfit and the idea of going into space."

Rocío's parents came from very different backgrounds: her father grew

◀ While there have been many challenges to her path,
Rocío's positive attitude keeps her going.

up in a family without a lot of money, where most people had not received a high school diploma, while her mother grew up in a community where the expectation was to attend college and graduate school and become a lawyer or a doctor or an engineer. But they both agreed that education was the path that would take their children further. Rocío says they sacrificed a lot to make sure she went to excellent schools, preparing her for a professional life. Rocío had never met any scientists as a child, but she was encouraged by her parents every step of the way.

SCHOOL DAYS

Teachers in high school were the catalysts for Rocío's interest in pursuing science as a career. She still remembers a physics teacher: "Yanira was so impressive! She was tall, and she seemed like she knew everything; she had such an air of knowledge and confidence. I wanted to know a subject like Yanira did, and be able to teach it to others." Another teacher, who taught environmental science, also inspired Rocío, though in a different way. When Rocío met that teacher, Betty, they didn't get along. Rocío struggled and refused to do her homework for Betty's class. But, Rocío realized, "I was only resisting that class because I loved the subject so much and I couldn't imagine how I would make a career of it, how I would make money doing it. But that teacher—she was the first to really plant the seed that this was an area I should at least consider."

Rocío decided that environmental engineering would be a subject that would allow her to pursue her interest in the environment and giving back, while also building a career that would honor her parents' efforts. She worked hard to get into one of the most prestigious university programs in Peru, and had just begun her classes there, when her family got the news: they had been approved to emigrate to the United States. "It was a painful process," Rocío says. "Going to that university and for that degree was my dream, and I had been working for it for five years. And I had to leave it."

Moving to the US was challenging. Rocío did not speak English, and the costs of college were beyond daunting. Since she was still only seventeen, she decided to re-enroll in a local high school to learn English and hopefully create a new path to university. She says, "Science was the one thing that got me through. I had decided, in my mind, that someday I would get a PhD. . . . I don't know why I was so focused on that. I didn't know exactly what I would do with a PhD, other than it meant I would be a real expert, have the tools and authority to do the work I wanted to. So I kept that in mind as I went back to high school."

Rocío was committed to staying close to her family, even after she graduated high school, and she did not want to take on student loans. So she began taking geology classes at a community college before eventually switching to a four-year university. She worked two or three (or more) jobs while taking classes, and says that one of the hardest parts was that she had very little time and energy to make connections, to meet other people interested in science.

Eventually, at the bigger university, she started to meet people and get opportunities for jobs and internships, but it took a long time. One of the professors she worked with at the United States Geological Survey (USGS) was an alumnus of Brown University, a world-class school in Rhode Island. He became a mentor.

Rocío laughs. "He was a white man from the USA, so nothing like me, but I thought, I want to be like him and do the work he does. I want to be a female version of him. I was pretty confident in my abilities and perseverance, but because he also believed in me and thought that I belonged at Brown, I had that extra bit of confidence."

Rocío worked at the USGS on micropaleontology, which is the study of tiny fossils one can only see with a microscope. For her work, she processed core samples of mud from deep in the sea to, as she put it, "travel back in time, like a geo-detective." Rocío studied the mud under a microscope to observe remains of microscopic animals and plants. She then analyzed their bodies and chemistry to figure out what life was like in different ecosystems when they were alive. This information allowed her to understand how different parts of the climate system interacted to create the record they

"I describe myself as a time traveler on Earth, where I get to use parts of the Earth to almost recreate the history of what happened on the planet at some point in time."

left behind. By studying past climate and the various factors that influenced it to change, from greenhouse gases to the volume of ice at the poles, scientists are able to better understand past and even present climate variation and what the future might hold.

"I describe myself as a time traveler on Earth, where I get to use parts of the Earth to almost recreate the history of what happened on the planet at some point in time."

She says, "I love micropaleontology. It's incredible to look at something the size of a sand grain or smaller, to learn about climate. To look at communities of microfossils and how they changed in time to infer how climate changed during a specific period of time. To play detective with these microscopic bugs and figure out what else we can learn from them." Rocío spent her time in graduate school learning to use more tools that would complement micropaleontology and help her play climate detective. She says, "I knew I wanted the micropaleo component, but I wanted to add other skills, whether through physics or chemistry, that would help me decipher more information and understand the bigger picture."

She says, "I loved collaborating with other people, nourishing my brain by learning so many things. I also loved sharing my love of microfossils, with kids and grown-ups alike. They are some of my favorite organisms, and I loved sharing my knowledge of them and what they mean to science."

The work was grueling for many reasons, and while Rocío felt welcomed at Brown and loved her time there, she also acknowledges that it was often challenging. She says, "I knew that I had things to offer, but I also went to a college that didn't prepare me well enough for graduate school at such a renowned university. I was working with people who spoke so intelligently on every single subject and even outside of the academic space. None of the professors, none of my peers but one looked like me. . . . There were times I felt like I wasn't sure I was supposed to be there, but I just kept pushing."

While Rocío loved the science, and the boat trips to collect samples, and the teaching, the pace was too much. She had gotten married, and after a couple of years of working on research and teaching at a local school, her health started deteriorating. It was at this point that she received a diagnosis of Myasthenia gravis. This is a chronic illness that prevents muscles from working properly and is highly variable. While Rocío had been battling symptoms for a while, the disease had reached a point where she knew she had to change her lifestyle, which, she admits, was often pushing her to the limit. Even defending her proposal to become a PhD candidate, a huge milestone of presenting publicly after all her studying, took a dramatic turn. She collapsed on the floor before it began. Rocío's determination shines through, though. She says, "I am the kind of person

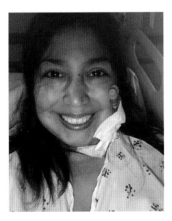

A selfie Rocío took from her hospital bed

"I was a new person after the diagnosis. I had to figure out who I was, both professionally and personally, after that."

ADVICE FOR YOUNG PEOPLE

The work of GeoLatinas is vitally important for Rocío, who recognizes what she missed in her own journey. "One thing I want kids to know, and one of the reasons I am doing the work I do right now, is that it was really hard for me. I learned so many things the hard way because I had lots of loving support, but very few mentors or people who could help me with this path. I wasn't looking for a community, because I had my family, but it would have been better and easier if I had found that nourishment along the way."

who is always looking for a different perspective, many times looking on the bright side, so I just got up and thought, well, that shook off the nerves, at least! And it worked out.

"I was a new person after the diagnosis. I had to figure out who I was, both professionally and personally, after that."

As a mother of biracial children, as a scientist and a person with chronic illness, Rocío thought a lot about how best to spend her energy and time. This is why she pivoted from her science research, instead pushing forward with her other constant goal of giving back. Rocío is the cocreator of GeoLatinas, an organization that supports Latina earth and planetary scientists. She is also the director of the CycloCohort Program, which supports scientists, particularly from marginalized backgrounds, who are just starting out in their career. Rocío's work still keeps her involved with science, but for the moment her focus is on mentoring early career scientists, coaching academics, and supporting womxn and Latina geoscientists, and she's happy with that.

She hopes that her Black-Latinx children, as well as all kids, will see themselves in science. "Kids have magic—they can ask questions and make something important just by asking." With the growing urgency of climate change, Rocío wants all people to get involved. "The first thing to realize," she says, "is that the Earth will survive, whatever happens. But that doesn't mean humans will, unless we do something. The rate of change, the way the environment is changing, it is all happening much faster than ever before. Climate change is linked to everything. It is linked to racism, and social justice, and equity: Who gets the worst of it? Who gets to make decisions? Who do we see on the internet? Who gets to be heard? I would urge all kids to remember that changes will need to be big to overcome this issue, but also everybody has a role to play, something to contribute. We need everyone's grain of sand."

She continues, "I would tell kids: Keep your curiosity, keep asking questions, and know that the necessary change is up to all of us, both individually and as a larger community. We can still make a difference."

CHRONIC ILLNESS AND INVISIBLE DISABILITY

For Rocío, getting a diagnosis of Myasthenia gravis changed her life. But what does it mean to have a chronic illness? For millions of people, it is a way of life. Having a chronic illness means the disease might get better or worse, but doesn't get cured. Many chronic illnesses are well-known, like diabetes or some forms of cancer. But many others, like Rocío's, can be unusual and hard to diagnose. And they can also be invisible. An invisible illness can also mean an invisible disability.

A highly variable, invisible chronic illness is one that is not outwardly noticeable to other people, and sometimes is even hard for doctors to identify and diagnose. And an invisible disability is having a mental or physical challenge that isn't immediately obvious to people around you. The challenge of living with an invisible illness goes beyond the symptoms. Often people feel that their peers and friends don't believe there's anything wrong with them, and that they just need to toughen up and get back to work. People with invisible chronic illness often feel misunderstood, and can even doubt themselves. Getting a proper diagnosis can make a real difference!

Science is a field where the expectation often is that people should push themselves hard, valuing their research above everything else. But with a highly variable chronic illness or invisible disability this can be impossible. One of the ways that more people can find themselves in science is if the scientific community recognizes that it's okay for someone to take more breaks, or work at a different pace, and give room to other things in life besides science. Hopefully that change is coming.

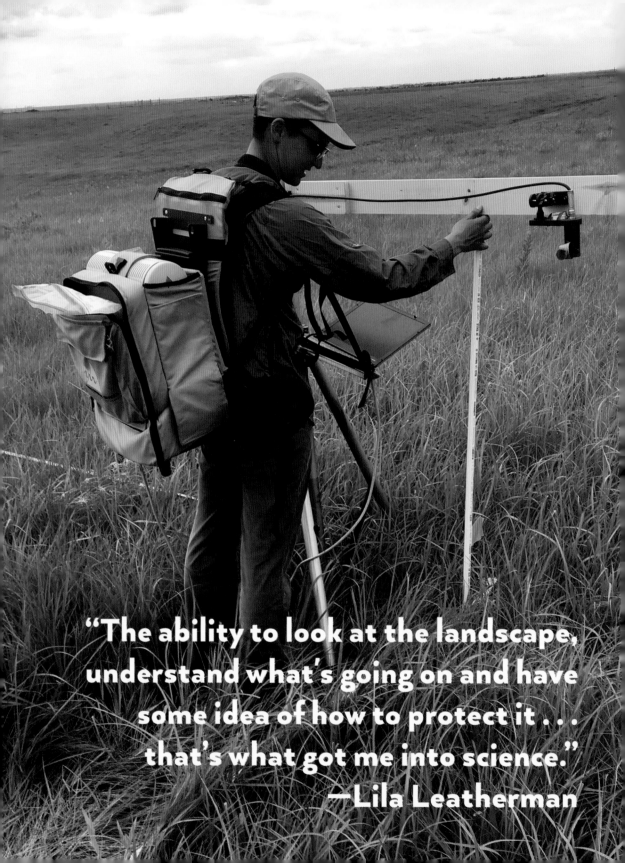

"The ability to look at the landscape, understand what's going on and have some idea of how to protect it . . . that's what got me into science."
—Lila Leatherman

LILA LEATHERMAN

Lila Leatherman studies dead needles on pine trees in Montana, or streams and rivers in remote corners of Alaska, all while sitting in their home office in Utah. Using satellite imaging techniques called **remote sensing**, they are able to focus in on wilderness areas, some that are almost impossible to get to in person, and take pictures that help inform scientists about the health of the land, water, trees, and other plants growing there.

"The ability to look at the landscape, understand what's going on and have some idea of how to protect it . . . that's what got me into science. And that is what has carried me through the whole way."

◀ Lila is using a remote sensing tool called an ASD field spectroradiometer to take images of grass at the Konza Prairie Biological Station in Kansas. Lila will compare the information gathered here with other images of plants in the area.

WHAT IS REMOTE SENSING?

Many scientists, including Lila and others in this book, rely on satellite images and remote sensing in their day-to-day work. **Remote sensing** is the process of finding and tracking the physical characteristics of an area remotely—from a distance—using sensors on cameras attached to drones, airplanes, or satellites.

While there are rumors of cameras mounted on pigeons in the early 1900s, remote sensing officially began with cameras mounted on airplanes in the 1930s. Those early aerial photographs offered scientists insights about the environment that they couldn't see from the ground. The sensors mounted on satellites today are much more sophisticated, capturing not just what our eyes could see if we flew above distant landscapes, but changes in electromagnetic energy, or light, beyond the visible spectrum. Different sensors capture images at different **wavelengths**, which refers to the part of the electromagnetic spectrum that a sensor is able to detect. Radar sensors, which detect wavelengths much longer than visible light, are also used in remote sensing! Radar sensors detect wavelengths long enough that they can "see" through trees and clouds to look at the shape of the Earth's surface itself. On the other end of the spectrum, X-rays have much shorter wavelengths than visible light, and can see inside the human body!

Depending on the way the photo is taken, the images might look more like photographs we take with our phones or like something very different—for instance, images that can help scientists understand the temperature and moisture levels of a particular area. This is important because sometimes just looking at a photograph—no matter how detailed and impressive—won't tell the whole story.

Remote sensing can help scientists understand in real time the impact of forest fires, invasive species, deforestation, and more. People like Lila are part of a global force who combine the power of technology and the power of human communication to better understand the threats to our forests. They use satellites and other tools to capture data all around the world, sharing the information with communities, governments, and businesses. Because while not everyone can get themselves into these remote forests, everyone needs them to keep our planet healthy.

"In my science classes, no one ever taught the fun part of getting to ask cool questions and walk around the world trying to understand what's going on around us."

When talking about how they got into this work, Lila says, "I had the good fortune to grow up surrounded by nature. There were fields to run in, bugs to catch, and creeks to play in. I wasn't thinking about it as science, but I was collecting worms and caterpillars, studying snails, and generally exploring the natural world around me.

"In my science classes, no one ever taught the fun part of getting to ask cool questions and walk around the world trying to understand what's going on around us."

Lila continues, "I was a good student in school, but I didn't connect with the science classes. It was a very strict, boring focus on process." Lila wishes that their teachers had taught students how to ask good questions and be curious about the answers, because as far as they're concerned, that's what makes a scientist. Instead, science was taught in a rigid and absolute way, underscoring what they think is one of the greatest problems in scientific study, from elementary school to university: "Science is too often painted as this model of objectivity, when of course objectivity is a myth." In a 2019 article for the website Massive Science, Lila writes, "Science . . . does not happen in a vacuum—science is part and product of a society."

WHAT IS BIAS IN SCIENCE?

We often talk about things being "objective" or "subjective," with opinions considered to be subjective (Coffee ice cream is the best flavor! Everyone knows cats are sneaky!) and hard facts considered to be objective (Water turns to ice when it reaches a certain temperature. Objects fall to the ground because of gravity.). Seems pretty simple, right?

Science is supposed to be objective, meaning it is based on research and provable facts, not squishy opinions. But the truth is that it's very challenging to be completely objective. Our **bias**, or our opinion or attitude toward something, influences us in all kinds of ways. And scientists are just as biased as the rest of us! The result is that "objective" science is more influenced by people's opinions than we realize. And when you look through history at who has been doing most of the science experiments and coming up with the results, you see that it is largely one group: American and European white men. This doesn't mean all their science is wrong—not by a long shot! But it does mean information that has been accepted as objective fact just might be more subjective than people realize.

This same bias of thinking that white men are the most suited to science is still far too common, and can make scientists like Lila feel unwelcome. One way that science is biased, Lila says, is that academic and research communities are largely built for the comfort and success of a certain kind of person, specifically white, able-bodied men. Because for a long time these men were the vast majority of scientists, they began to assume that their definitions of "average" or "normal" or "typical" were correct for all people. These assumptions can range from the fact that laboratories are set up in ways that make them challenging for scientists with physical disabilities, to expectations about communication style or even how people dress. The result is that the so-called objective norm is uncomfortable or unwelcoming for many LGBTQ+, disabled, and non-white scientists often enough that they decide to leave the field.

And in case you're wondering how real this is, a group of—you guessed it!—scientists did an experiment to test bias. They created a group of made-up candidates applying for jobs in science, all with identical résumés. These imaginary candidates went to the same schools, they had the same jobs, they had all the same details. The only difference? The experimenters gave some of these imaginary candidates names people associate with women (Jennifer!), and some of them names people associate with men (Kevin!). Remember, these made-up people had the exact same qualifications! But guess which ones were rated as more capable of doing the job, and offered higher salaries? If you guessed the candidates who appeared to be men, you'd be right.

And a similar study was done years later, where the imaginary candidates not only had different masculine and feminine names, but some also had names that suggested they were not white, but instead were Asian, Black, or Latinx. Once again, the people who were hiring these imaginary candidates chose the white men first. So there it is: it's not a subjective opinion that there's bias in science, but actual objective fact!

Why does this matter? Does this mean science is fake and shouldn't be trusted? Not at all! But it's important because we all have our own biases, and part of being a good scientist is noticing and paying attention to all the factors that might influence your research. Good scientists, and good humans, have to question whether their information is based on fact or opinion. And we have to remember that any version of "typical" or "normal" that makes some people feel unwelcome or uncomfortable is, objectively speaking, not good enough.

While Lila had enjoyed nature and the outdoors as a kid, they did not plan on studying science in college. They say, "I started college as a straight English major, and graduated as a gay scientist!" They took an environmental-studies class as an elective, and realized that this was what they wanted to study. Lila says, "It was a moment of realizing I can look at a tree, and be able to explain how it's growing and how the water is moving from the roots out to the tips of the leaves, and how the wood is connected and constructed.

Even as a kid, Lila loved being outdoors.

"You mean I can have all these skills to help me understand the natural world around me, and the tools to help protect it? That was it! That was for me."

"You mean I can have all these skills to help me understand the natural world around me, and the tools to help protect it? That was it! That was for me."

Part of what drew them into science, Lila says, is the professor who taught it. "She was a rad woman who everyone wanted to be around, and that was a huge part of the excitement."

They continue, "I took a class where I learned how to identify hundreds of species of native plants in Ohio, where I was at school, and I was hooked. I was able to walk around and it felt like I was saying hello to my friends; every spring the same flowers would return. Being able to have that intimate connection with a landscape was really important to me."

After graduating from college with a degree in biology, Lila went on to pursue their PhD, though they decided to leave the program after completing three years of the course work. While they liked some of the work, in retrospect Lila says, "In the program I was in, there were a lot of ecosystem questions on a global scale, and I felt that my interest in what was happening on a local level was considered unimportant."

Lila warns against the focus, in science, on prioritizing what is large and flashy and will get headlines, above projects that might be smaller in scale

Lila greeting their floral friends

"I decided pretty early that I would rather be a good person than a good scientist, and that having a robust life outside my job was more important than succumbing to pressure to get my work published."

but can benefit communities. They also recognize the challenges of being part of the LGBTQ+ community, and specifically coming out as non-binary and transgender while in the PhD program. While most people were not openly hostile, Lila says that almost no one remembered to use their pronouns, or sought to make them feel welcome and included. In the article for Massive Science, Lila writes, "Feeling unable to share your full identity at work is both exhausting and isolating—it's tiring to feel like you're living a double life, and challenging to find other people who can relate to the struggles you're experiencing."

"I decided pretty early that I would rather be a good person than a good scientist, and that having a robust life outside my job was more important than succumbing to pressure to get my work published."

Lila took a break from practicing science and spent a year exploring other interests, from rock climbing to educating people about the need to make outdoor spaces more welcoming and inclusive for transgender people. They enjoyed the work of partnering with outdoor expeditions and local climbing groups, but they also missed practicing science.

Lila returned to science and their beloved plants, and now uses satellites to make maps of plants and forests. Their work involves using images to understand how healthy or unhealthy forests are, and what factors, like insects or disease, might be impacting the trees. This work is in partnership with the US Forest Service, and they make maps and do geographical analysis across the country. Using their computer, Lila looks at forest health all over the United States.

Lila says, "It's so cool to have a little window into all these places, and to get to learn about all of them. I am now back in a job that uses many of the same skills and tools I was using in my PhD program." They continue, "I really like the work. I like telling stories and asking questions and connecting to people who work in the Forest Service all over the country." In their role Lila is often helping collect and understand the information that will allow the Forest Service to decide how best to help the land and the ecosystem.

"I am doing remote sensing, using satellite images to get information about the surface of the Earth, and figuring out how it has changed over time."

They say, "The power of the tools that I use is the ability to look at places you can't access off the highway or from an overlook, and to get a bigger picture of what's going on. And not just a bigger picture, but a longer-term picture of what has happened over time. I can gather information on a hundred-square-mile area every year for the last thirty-five years."

"I am doing remote sensing, using satellite images to get information about the surface of the Earth, and figuring out how it has changed over time."

Lila clarifies that they are not directly responsible for creating land management plans, but rather for using satellites and tools to understand the scale and scope of the problems the land might be facing. They say, "My job is to identify the threats. I can say that the grass in this area is less healthy than it was ten years ago, and that correlates to higher temperatures and less rainfall. Or I might notice an extreme decline in forest health in a certain area due to pine beetles." In the first scenario, Lila might recommend that the Forest Service plant a more resilient species of grass, and in the second one they might suggest a pest management plan. Once Lila has helped focus in on the problem, others come up with a specific plan to address it.

One of the elements of this work that Lila enjoys is the mix of technology and human interaction in understanding the landscape. They teach people in the Forest Service the basic concepts of the tools they're using, offering an overview of remote sensing. One of the things they love explaining is how satellites can use the electromagnetic spectrum to give us pictures that can tell all kinds of secrets about what's going on at the Earth's surface.

Lila explains that these specialists can then use the information to make informed decisions. Whether they involve pest management or new plantings, these decisions are largely based on information that cannot be gathered with boots on the ground, either because the land is too remote or because the problems can only be understood by looking at a longer time frame or a larger geographic area. One of the great things, Lila says, is that they can collaborate with people who might know the areas intimately, all from their computer.

When asked if their younger self would approve of the work they do, Lila laughs. "When I was a kid I wanted to be a cowgirl paleontologist studying dinosaurs in Montana. And I'm none of those things—I'm not a girl, and not in Montana, and not studying dinosaurs. But I live in a beautiful part of the world near the mountains in Utah, and I study plants, so I think I would have been happy with that!

When it comes to giving advice to other kids, especially transgender and nonbinary kids, Lila is clear. "It can be really hard, but there are people who will surprise you, and who will have your back when you don't expect it. And you must remember that you deserve to have respect and support. And it can be hard to ask for that. But you deserve it, and if you are somewhere where people don't respect or understand you, then you are allowed to quit and find someplace else."

"There are a lot of ways to do good science, so if the place you are is not the right place, you're allowed to try another road."

And Lila has one more message: "One important thing that all transgender and nonbinary kids should know: Science is on your side! It is not binary. There are fungi that have hundreds of different sexes. There are animals with different sexes that take on all kinds of roles within their social systems. There are animals that change sex over the course of their lives. There are plants that have both sexes. Science supports queer and intersex and trans and nonbinary people, even if the scientific community doesn't always recognize it. Always remember that nature is very, very queer!"

"There are a lot of ways to do good science, so if the place you are is not the right place, you're allowed to try another road."

"You have to be brave and follow your own heart."
—Marshall Shepherd

MARSHALL SHEPHERD

When Marshall was in sixth grade he decided to enter a school science contest by answering a question: The weather forecast predicted the weather in the large cities near him, but how could he predict the weather where he lived? Using everyday objects, Marshall built a weather station that did a pretty good job! (He also won the contest.)

Since that sixth grade science contest, Marshall has kept asking questions and studying weather. Today Marshall is a **meteorologist** and a famous expert on weather and climate change. He has worked for NASA, appeared at the White House, and spoken around the world about weather

◀ Meteorologists get asked about the weather a lot. While Marshall might not always know if it's going to rain, he's not likely to get caught without an umbrella.

and climate. Marshall is quick to point out that studying weather isn't the same thing as being a weather forecaster on TV. "I knew I didn't want to be on television; I wanted to study the whys and hows behind the weather."

WHAT IS METEOROLOGY?

Weather is all around us all the time, but what is it, really? Warm or cold, pouring rain or bright and sunny, pelting hailstones or fluffy snowflakes . . . it's all weather. The Earth is surrounded by layers of gases, called the atmosphere, which work to make the planet habitable by insulating it, cooling it, and creating conditions suitable for life. Whatever the temperature, clouds, and moisture are doing is related to the Earth's lower atmosphere, where we live. And all of it is studied by **meteorologists**. They study the weather, and in particular the physics and chemistry that create weather, using tools to observe and analyze what's happening in the atmosphere.

Meteorologists research weather patterns, collecting data and using the information to predict what might come next. They use tools like satellites, which can gather and share details about weather patterns across the planet, and computer programs, which can take huge amounts of data from satellites and help find patterns. Sometimes this is used for forecasting, like when a meteorologist on TV tells us if our weekend plans will be rained out. But this research also goes beyond the short term, to help with long-term planning on everything from building roads to managing farms to understanding human and animal migration patterns. Literally every aspect of our lives depends on the weather: the crops that grow into food we eat, the shipping of the stuff we buy across oceans and along highways, the safety of our communities when faced with storms or extreme weather. Governments, corporations, sports teams, towns . . . everyone who has to make decisions and plans uses the research that meteorologists do.

Because of their study of weather patterns and predictions, meteorologists have a front-row seat to our changing climate. As they study floods, droughts, tornadoes, extreme temperatures, and other weather events, they have seen that human activity is changing our climate in new and dangerous ways. Whether they are tracking tornadoes or assessing the conditions for wildfires, meteorologists witness the effects of our changing climate firsthand. And they are using this information to help people learn how we can take actions to better protect our planet and our communities.

"After that sixth grade project I was pretty sure I wanted to study weather, and even started thinking about colleges that had good meteorology departments."

Marshall knew from that early moment that his path was set. He says, "After that sixth grade project I was pretty sure I wanted to study weather, and even started thinking about colleges that had good meteorology departments. Now that I'm a dad, I know not all kids have any idea what they want to do at that age, but I was unusual in that way."

Marshall credits his mom for some of his focus. She was a fourth grade teacher and made sure that he understood how important education was. In an interview with the NASA Center for Climate Simulation, Marshall also gave a shout-out to his sixth grade teacher, who, he said, "first introduced me to the principles of the scientific method."

While there weren't a lot of meteorologists or atmospheric scientists in his community when he was growing up, his mother did make sure he had access to the library, where Marshall remembers reading everything he could find about science. "I was fascinated by Black scientist George Washington Carver, who wasn't an atmospheric scientist at all; he studied peanuts at Tuskegee University. But still, I read anything I could find about him."

While the weather is all around us all the time, most people don't really know what it means to study the weather. Ask people about meteorology, and most will mention the people on TV, but Marshall says that only around 8 percent of meteorologists are TV weather forecasters. And Marshall wants to change that misunderstanding. "I spend a lot of

"When you work in a global world, weather and climate change really matters."

time educating people on what meteorology is. I am the director of a major program at a university, and my graduates work at places like major airlines, or the National Weather Service, or federal agencies, or large private companies. When you work in a global world, weather and climate change really matters."

When asked what the hardest part of his journey to science was, Marshall is clear: "People really don't understand this field, so they don't realize how heavily it's based in calculus and physics and fluid dynamics and thermodynamics. It's rated one of the most difficult majors on a college campus." He acknowledges that as a professor, "one of the hardest parts is that I've had to break kids' hearts, students who walk into my office saying 'I love hurricanes and tornadoes and clouds,' but who don't realize the difficulty of the science."

Another challenge Marshall faces is that as a Black scientist he's in a field where Black people, as well as other underrepresented groups, are few and far between. He says, "There's always a whisper that I got into the rooms I'm in because I'm African American. I've had certain successes in my career. And I'm aware that those successes came from the levels I have reached. But you talk to any African American professional in a field like mine, and there are rumors. Someone will write in a blog that I'm a token Black person, or question my perspective on climate change while mentioning my race. . . . These kinds of things are not unusual."

He continues, "It would be naive to think this kind of behavior isn't common, but I have been able to ignore it for most of my life. Even in high school, when I did Model United Nations, or poultry judging with 4-H, I was often doing things my Black peers wouldn't do. But I was interested and curious, and never took the time to stop and see who else was around me, and if they looked like me. I was interested in what I was interested in."

Marshall had always wanted to work for the space agency NASA; when he gave a speech at his high school graduation he said that working there was his dream career. After he finished his degrees, that dream came true. He worked at NASA for several years and wasn't sure he should leave, even though he felt ready for a new challenge. After all, working at NASA was what he had always wanted from the time he was a teenager! Leaving seemed like a risky idea, but as he says, "When I left it opened up amazing opportunities. The lesson was clear: don't lock yourself into one trajectory too early, or be too rigid so that you avoid change. Because you never know what's next. You have to be brave and follow your own heart."

"You have to be brave and follow your own heart."

While working for NASA was a dream come true, and Marshall is frequently mentioned as a candidate for high-level positions in various presidential administrations, he says that he loves what he does now and isn't interested in making a change. "I am in a free and entrepreneurial environment where I can do research on whatever I want, and work with students and see them blossom, and write for magazines and newspapers and blogs." He goes on: "At my core, I'm still that sixth grade scientist. I enjoy asking a question and getting funding to go off and do research and discover new things."

He says, "Every day my job is different. Some days I am teaching and working with students. Others I might be more focused on my research and meeting with collaborators. Then still others I might be meeting with folks at NASA or in Congress on Capitol Hill. Then there are days when I might be doing a TV show, talking about weather and climate. There is no typical day."

"I believe that every experience I've had has had a lesson in it. I don't know that I would change anything."

Much of Marshall's own research is done sitting at a desk, not out in the weather. "I use complex models and satellite data sets, radar, and various other advanced tools to look at weather and climate change. Some of my colleagues are out in the field with instruments to measure weather on the ground, or flying from balloons. But my work is more using computers."

He continues, "The fundamental principles of my scientific research haven't changed that much from my sixth grade project. I use the scientific method: I pose a research question, find data and methods that can test my hypothesis, then I conduct the research and do the experiments, and finally I report on my findings. It's at a much higher level, and I'm using multibillion-dollar satellite systems instead of homemade weather instruments. And instead of writing up my science project report, I'm publishing my results in a peer-reviewed journal or speaking about it at a conference, or presenting to Congress. The scale is different but the fundamentals are the same.

"I believe that every experience I've had has had a lesson in it. I don't know that I would change anything."

That's not to say it's always been easy. While graduate school in the sciences is almost always something that people are paid to do, rather than having to pay for it (like college), staying in school for years and years can be difficult. And being a Black scientist has its own challenges. Marshall tells of microaggressions, moments of disrespect or exclusion due to his race. And he tries to help his own students with that. "No one told me to be on the watch for that kind of behavior," he says, "and I had to learn how to deal

ADVICE FOR YOUNG PEOPLE

When asked what advice he has for other kids who might be interested in science, Marshall's response is surprising from someone who knew what he wanted to do from age ten! He says, "From an early age I was fascinated by the concept of a Renaissance man, a person who knows a lot about many different diverse subjects. So even though I was interested in weather, I did Model United Nations, I was in 4-H, I played sports, I read. So my advice is: find your interest, but don't get too narrowly locked into that. Because some of the things that have helped me in my career, like my communication skills and my writing skills, didn't come from my training as a scientist."

He also stresses the importance of mentors. "Find people you can reach out to, even from afar. For me it started with reading books about people and realizing that there were people who had gone before me. One advantage kids have today is that email and social media can help you connect with people! I encourage kids to do that—just email someone, and ask questions. And it's okay to change your mind; even if you do approach someone, you might not wind up staying interested in that area. But it's still good to get those answers."

with it by myself. Now I pass some of this knowledge on to my students, so they can prepare for it."

When asked what some of the weirdest parts of his work are, Marshall says that a lot of his research is on urban climates. "And did you know that cities like Atlanta or Houston can cause their own thunderstorms? It's true. Cities can make their own weather."

He goes on: "To understand how the city of Atlanta can create or modify its rainfall, I used a tool called the Global Precipitation Measurement, or GPM. I helped develop it when I was at NASA, and it's a satellite system that can measure rainfall from space. Whereas fifty years ago I would have had to travel to the city and put physical rain gauges [devices that collect rain for measurement] around the area, then set up a network of observations on the ground to measure rainfall, now I can use this sophisticated tool to get images of rainfall every few hours, over years! So I can build up an incredible research database that way."

WELCOME TO HEAT ISLAND

While traveling to an island might sound like fun, **heat islands** aren't exactly a vacation. A heat island is what happens when an urban environment traps the heat and makes it much hotter than nearby rural areas. This happens because of a few factors. The first is that dark colors trap the heat, so all the concrete, pavement, and asphalt in cities really heat things up. In addition, the lack of plants and trees in cities plays a big role. Plants have a way of bringing water up from the earth and pushing it out through tiny holes under their leaves, which in turn helps bring down the overall temperature. It's nature's way of turning on the air conditioner, and without it, cities get hotter and hotter!

And the heat is only the first step. Scientists have learned that all that hot air in cities actually causes more clouds to form over urban areas than rural ones. So as Marshall says, big cities might build their own thunderstorms . . . which at least might help people cool down!

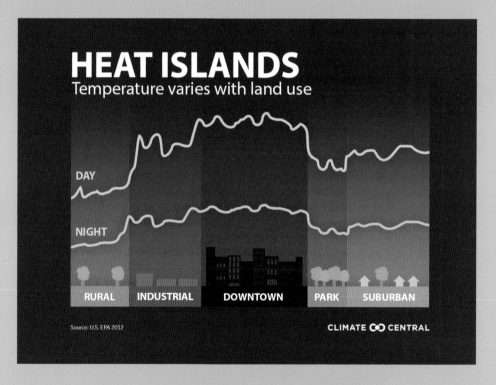

This infographic on heat islands demonstrates how daytime and nighttime temperatures in a location can vary depending on what the land is used for, from rural to suburban areas.

"My interest in climate change often focuses on how the changing weather impacts vulnerable populations."

Marshall's work on weather is linked to the Earth's changing climate. "My interest in climate change often focuses on how the changing weather impacts vulnerable populations. I look at how certain groups of people might have to be evacuated, and how climate is impacting access to everything from jobs to agriculture."

In his home state of Georgia, he has done research on a climate vulnerability index, looking at factors for every county in Georgia to understand which are most vulnerable to climate change. He points out that in some cases environmental factors like flood areas are what make a community vulnerable, but in other cases the risk is from societal factors, like poverty levels and whether people have access to health care. All of these elements can impact how climate change can hurt a community.

The other part of climate science that Marshall works on is the communications side. "Scientists can be really bad at talking about important subjects to non-scientists! Sometimes we tend to use lots of jargon or statistics, instead of explaining it in ways people can understand. So I spend a lot of time trying to communicate to people."

When it comes to what people can do in their own homes to fight climate change, Marshall is clear. "Climate change is not something we're going to solve by people changing their light bulbs. The global climate crisis is going to be solved by large-scale transformational solutions that involve the reduction of carbon emissions and changes in land use. The Paris climate agreement is a good step in that direction.

"But that doesn't mean that individuals can't do anything! We've been studying how to draw down carbon emissions, and one solution that has emerged is **composting**! If we start in our own homes, composting food waste in our backyards and turning that into soil, that reduces food waste in landfills, which becomes methane, which is a greenhouse gas. All of us are part of the collective discussion, so for students to talk to their parents and their neighbors, that creates a broader narrative. If you make people around you more aware, maybe they'll be more likely to vote for candidates who will help, or take action of their own."

Marshall's own work on urban climate keeps him busy, but when asked what's the most exciting science happening in his field, he mentions another area entirely: artificial intelligence. "Some of my colleagues are taking data from our observation and our models and feeding it into artificial-intelligence and machine-learning programs. These programs

ATLANTA
THE POWER OF TREES

25.1 MILLION TONS — CO₂ EQUIVALENT REMOVED

16,912 MILLION GALLONS — STORM RUNOFF AVOIDED

544 MILLION POUNDS — AIR POLLUTION ABSORBED

Source: U.S. Forest Service i-Tree County Tool

CLIMATE CENTRAL

Trees can provide a natural air conditioner effect, making leafy suburbs and rural areas measurably cooler than cities. Neighborhoods that mostly house Black and immigrant communities often have fewer trees and green spaces, making them hotter.

WHAT IS COMPOST?

Compost is made up of organic matter that has decomposed. Literally anything biodegradable can turn into compost, from food scraps to raked-up leaves to old newspapers. There are commercial compost companies that collect household and restaurant scraps and turn them into compost in industrial machines. But there are also backyard compost options, where you just collect certain biodegradable waste and let it turn to useful garden gold! (There are more specific rules about what can and can't be composted in a backyard compost pile versus a commercial compost company, so check your town's rules for guidance.)

Compost can be used to fertilize gardens and is considered incredibly helpful by gardeners and farmers. The process of making compost is called composting (clever, right?), and it's the process of breaking down organic matter. And not only is compost super helpful for growing new things, it also keeps literally millions of pounds of material out of landfills every year! According to the Environmental Protection Agency, around 30 percent of the food we throw out could be composted. Think of it like recycling, but for a banana peel . . . it gets turned back into soil to grow more food.

can start to develop systems where they know exactly how much water to use to irrigate a crop without wasting any. Or airplanes may be able to know, based on what kind of atmospheric patterns are in place, how much turbulence to expect."

Interestingly, the other area that he finds fascinating is not nearly as high-tech as artificial intelligence. It's the question of how people consume information, and why they react to certain warnings but not others. As humans face more impacts from the changing climate, scientists like Marshall are studying the science of what helps people listen and take action. As he says, "We keep breaking records—for the hottest year, for the most rainfall, for the most hurricanes. This is the new normal. So now the question is, what are we doing to do about it? What are our solutions?"

Marshall is the first to admit we don't have all the answers, but he's optimistic. "I think people do care. We aren't going to solve this quickly, but we'll start chipping away."

"I was fascinated by what I now know is natural history."
—Daniel Palacios

DANIEL PALACIOS

Daniel Palacios, a professor at the Oregon State University Marine Mammal Institute, grew up in Bogotá, Colombia, a landlocked city of seven million people in South America. He is the son of first-generation college graduates who valued education and bought him encyclopedias and science books when he was a child. "There were books about planets and interstellar voyages, but that didn't interest me much. I was always drawn to the volumes about horses and dogs and African animals and prehistoric animals. I was fascinated by what I now know is natural history: the diversity and variety of animals and how they have different habitats and ways of evolving. But at

◀ Daniel's fascination with the natural world and marine life have led him from landlocked Bogotá, Colombia, to the edge of the Pacific Ocean in Oregon, United States.

"It was such a dramatic decision, to move away from everyone I knew, and from the only place I knew, but I decided to do it."

the time I didn't have a name for it. I was living in a very urban environment, and completely drawn to the wild animals."

Daniel's parents sacrificed a lot to prioritize his education. He went to a very challenging and strict religious school and was expected to go to college. He says, "I grew up completely surrounded by urban occupations, like doctors and lawyers and engineers. Those were the kinds of successful careers my parents imagined for me. I didn't know any biologists, or really anyone who did science at all. And by the time I was fifteen or sixteen and thinking about college, I knew those other city careers were not for me."

While his parents were worried about Daniel's lack of interest in traditionally successful paths, they weren't discouraging. "They wanted me to be able to make a living, but at the same time they encouraged me to pursue something I was interested in." He continues, "I think it was hard for them to fathom my interest in science, since it was something they had no exposure to."

As he thought about what he might study, he wasn't sure what kind of college to pick or what to study. He was still fascinated with animals, but had no interest in being a veterinarian. He considered marine biology as the next choice. But there was only one college in the whole country that offered a marine biology degree, and it was far away.

Daniel says, "It was such a dramatic decision, to move away from everyone I knew, and from the only place I knew, but I decided to do it."

Daniel credits his high school for helping him prepare for the work at college, and recognizes his parents' role. "My parents grew up in a humble environment, and they put us in the best schools they could afford, because they really believed education was the most important thing. And because I went to such a challenging school, I was very well prepared for college. It made me feel more confident." Because he was so well prepared, Daniel had more free time and used it to talk with his professors, learning about what older students and researchers were doing. It was exciting for him to learn that research on whales and dolphins was happening right there in his own country. Even as a new college student, Daniel was starting to gather information about what was possible.

Daniel is clear that it was his interest and willingness to approach people and talk to them that allowed him to build his career. It wasn't until he met people doing research that he realized "people were doing this work and working with animals, not just studying them in books! It was the first time I saw a possibility that I could do this." He kept asking questions, and kept looking for ways to get involved. "I met someone studying the pink river dolphins in the Amazon, and asked if I could go with him on one of his trips. And it went from there."

> ## "My parents grew up in a very humble environment, and they put us in the best schools they could afford, because they really believed education was the most important thing."

"As a gay kid at an all-boys school, it was very oppressive. There was a lot of bullying. So getting to college, where there were men and women, and people were more open-minded, helped me find my own group of great friends."

There were other ways college felt like a great fit for Daniel. He had realized, from a very young age, that he was gay. He says, "I knew as a small child I was different," but also that, in the strict Catholic country of Colombia, "being gay could bring shame and guilt." He continues, "As a gay kid at an all-boys school, it was very oppressive. There was a lot of bullying. So getting to college, where there were men and women, and people were more open-minded, helped me find my own group of great friends."

He describes the early days of college: "It was like a buffet, and I could see everything and try different possibilities." But then, he says, "You start dealing with the realities of grown-up employment, and building a career. And there are areas of science that get much more funding, and that have more resources, but there aren't a lot of opportunities to devote yourself to studying whales and dolphins. So I had to make a tough choice. But I decided I wanted to study these animals, these fabulous, almost mythical animals."

The decision was particularly challenging in Colombia, where, Daniel says, there were few opportunities for him to continue his studies. He knew that he would need a graduate degree, and that he would need to leave the country. He applied to programs in the United States, among other countries, but he says it was difficult. There was no one at his university who was able to help him with the process, and, as he says, "It was challenging to understand the system and to figure out funding opportunities. I asked as many people as possible for information, but many were discouraging . . . not about the science, but about the process and the possibility of finding an opportunity to study abroad."

But Daniel was tenacious. As he sent in applications and tried to figure out his next step, he learned that a research vessel with a well-known American researcher and an international team of scientists was sailing down to the Galápagos Islands for a one-year expedition and would be sailing through the Panama Canal. A scientist Daniel knew was friends with the head researcher and was planning to meet him in Panama, just for a visit. Daniel begged to come along and be introduced.

He says, "It was wishful thinking to think this research vessel would bring me with them. But I was twenty-two, I had an aunt I could stay with in Panama, and I had no other plans. I figured it was worth it. There was a chance they'd take me to the Galápagos."

When they got to Panama, Daniel was introduced to the head researcher, who told him that for the next week the crew wasn't doing any science, they were repairing the ship to ensure it was seaworthy for the passage to the Galápagos. Daniel laughs as he remembers what they told him. "They said, 'Right now we are scrubbing and painting and repairing the boat. But if you want to help with that, and hang out with us, you're welcome.'"

He goes on, "I was there for a week, and every day I showed up and helped and became friendly with them. And on the last day they said, 'Okay, you can come with us, but when we arrive in the Galápagos you have to get off the boat, and you have to make your own way home.' And of course I agreed."

But once again, circumstances—and Daniel's efforts—led to another opportunity. When the ship arrived in the Galápagos there were some problems with the various permits from Ecuadorian authorities that they needed to conduct their research. Daniel, who spoke Spanish, was able to help sort out some of the challenges. He wound up staying with them and helping with the research for a year.

"It was amazing," he says. "I got a lot of experience, and met some very important people who helped me later in my career. Also, since I proved myself, and made myself an essential part of the crew, they invited me to come to the lab in Massachusetts to continue the work. I was still trying to get accepted and get funding for graduate school, so this was a perfect next step. At least I would be in the United States."

Daniel with his mentor, Dr. Bruce Mate. They first met on the fateful boat trip to the Galápagos in 1994, when Dr. Mate came to use the boat as a platform to tag whales for his work at Oregon State University.

"At each one of those steps I felt like I was literally walking to the edge of a cliff and jumping off."

Looking back, it can seem like Daniel had a clear path: college, then work on an acclaimed research vessel, then graduate school. But at the time, it felt very different. "At each one of those steps I felt like I was literally walking to the edge of a cliff and jumping off. When I went to Panama I thought I was crazy, but I figured, I have my aunt to stay with if it doesn't work out. And then when they invited me to the United States I had no plan for graduate school, but thought that at least this is similar to what I'm looking for . . . I should do it. But each time I was jumping without knowing where I would land."

He stresses that "there is a steep learning curve, and I wish I had known more about what it takes to be a successful scientist, because it's a narrowing funnel, and you have to take the right steps to get there. And if you have missteps along the way it makes it harder. Looking back, I see how I did things that made it harder."

Still, Daniel did get into a graduate program in the United States, with funding to pursue his PhD. When asked what he likes best about his job, he says, "It's about discovery and finding patterns in nature. When I was young and interested in natural history, I was doing the same thing, on a different scale."

Today Daniel has a PhD in oceanography, and is a lead researcher at the Whale Habitat, Ecology, and Telemetry Lab at Oregon State University's Marine Mammal Institute. While he doesn't get to spend his days hanging out up close with whales, his life's work is researching, tracking, and studying these creatures that have fascinated him since he was a kid. He uses telemetry, tagging whales with small devices that allow the lab to track their migration patterns, movements, engagements with other whales, and more. By learning what happens to these mysterious giants when they disappear beneath the waves, scientists are better able to understand how to protect them. For instance, they know that ship traffic, with its noise and sonar disruptions, can alter whale behavior. So knowing where whales travel to feed, mate, or raise their young can help scientists make the case for protecting those areas.

While many whale species were close to extinction years ago when they were hunted for their meat and their blubber, most are rebounding to healthy populations as international laws have been put in place to protect them. Now, however, whales all over the world face a new threat, and it's the same one we are all facing: the negative impact of human behavior on our planet. In an article for *Discover Our Coast* magazine, Daniel says, "Today the biggest challenges to whale conservation are largely the same ones that affect marine ecosystems as a whole: chemical and noise pollution, shipping, habitat degradation, and overharvesting of marine resources for human consumption." These challenges require everyone—not just whale hunters—to pay attention and change their behavior. And since whales travel huge distances and respect no borders, they also require international cooperation among scientists, governments, and citizens to help them.

And Daniel is quick to point out that helping whales helps the planet, which helps everyone. Believe it or not, there is even a way to put a price tag on the value of a wild whale: their contribution to helping fight climate change is worth around two million dollars per whale over its lifetime, according to the International Monetary Fund, a global organization that focuses on international cooperation about money.

WHY IS A WHALE WORTH TWO MILLION BUCKS?

Most of us would think about donating money to help organizations that are saving the whales as a charity, something that we would do to be a good person. But maybe it's actually more of an investment! How much would we have to pay to try to fix some of the ways we are causing climate change? Turns out whales do this work for us, for free!

One of the biggest challenges we face in the fight against climate change is capturing the carbon that gets released into the air through human activities and pollution. Forests, trees, and plants are excellent at capturing carbon and helping our planet. And it turns out, so are whales! And they do it all in the most natural way possible: by pooping massive whale-sized poops into the ocean that release all the nutrients from the tiny marine organisms they've eaten. The nutrients in their giant poops help the growth of marine plants that capture carbon, just like fertilizer applied to plants on land.

Whales live a long time, but when they die of natural causes in the ocean, they help the planet even more. All their lives they store carbon in their massive bodies, and when they die, they simply sink to the ocean floor. The carbon sinks with them, and stays down there long after their bodies are gone.

When we hear about science contests with million-dollar prizes for the inventor who can come up with new technology to capture carbon and fight climate change, it's important to remember: sometimes we don't need new technology! Sometimes we have to protect nature and get out of the way, so that animals like whales can do their job just by eating, pooping, and eventually dying.

Daniel points out that his work with whales is still based on observation of animals in nature, only he's not using his eyes. The technology he uses allows him and his team, along with colleagues in other countries, to truly understand the way the humpback whales near Oregon travel and live. And by understanding them, the team can better protect them.

"I'm observing nature with advanced tools that are collecting high-resolution data. This data gets analyzed by computer, and we often need years worth of information before we can piece together some patterns. But then we can make discoveries! We find things that are not immediately apparent, and then ask more questions. And look to find more patterns. That's still the most exciting part of my work."

"Keep the passion alive for whatever it is you're drawn to. But at the same time, in order to pursue a career in science, pay attention to other parts of your education that might not seem as important."

ADVICE FOR YOUNG PEOPLE

When asked what advice he would give kids interested in science, Daniel says, "Keep the passion alive for whatever it is you're drawn to. But at the same time, in order to pursue a career in science, pay attention to other parts of your education that might not seem as important. If you're passionate about animals, you might not care a lot about algebra or trigonometry, or computer science and physics. But these are essential foundations. And they are the ticket to be able to continue to study what you're passionate about." He recognizes that "for some people these aspects will be boring and difficult, and you can't see why it matters. But the good news is that once you get past these building blocks you will use these tools to answer the questions you care about. Science requires a certain language of math and statistics, and you need to make sure you can speak that language, because that will open doors. If you come from a background that's less common in science, people judge you more. So it's important to have these skills, because they will let you do the work you're passionate about."

Daniel points out that whales are difficult to study, especially for any length of time, because they spend so much of their lives deep under the water. For a long time, researchers were stuck sitting in boats or on shore, hoping to see them. He says, "They were truly a mystery. But now we can tag several whales at the same time, using small tracking devices that send data back to us, and see what the whole population is doing. We track whales using satellites, and we can use a device to continue to track a whale for years, day after day. This is a long process. We track them for at least two or three months, then it takes a few years to draw our conclusions and publish the data."

◀ Daniel doesn't just study whales. In this photo, he is off the coast of Morro Bay, California, deploying a set of instruments known as C-PODS, which recognize and record the sounds produced by harbor porpoises and other dolphin species. Daniel states that these harbor porpoises are very small and shy, and scientists have to be creative when researching them.

"I need to live, function, and succeed in cultures that are sometimes radically different from the culture I grew up with."

Asked what he does all day, Daniel says that unlike what people think, "I'm not spending my time swimming with whales. They are so fast—if you're in the water with a whale it will go by you in a flash. I am out on a boat for a week or two, tagging whales, but then I am back at my desk, tracking them via satellite. As a professor at a large university I also spend a lot of time managing projects, writing grant proposals, and doing other organizational tasks. So there is the business of doing science, then the actual research, or analysis of the data, or writing up the findings."

The challenges still continue, even as an adult. Daniel says that while his path has afforded him mobility to live and travel internationally, it comes at a cost. "I need to live, function, and succeed in cultures that are sometimes radically different from the culture I grew up with. And while globalization has made this easier, there's always an element of discomfort and insecurity."

And being a gay man in science can be uncomfortable as well. Daniel says that when he was younger, "I didn't know other gay people, and there are no role models for gay scientists in my professional field." As a result he kept that part of himself hidden away. But now he is open about who he is. In his profile on the website 500 Queer Scientists, Daniel says, "Because homophobia, violence, and discrimination . . . are still rampant in our society, I now feel that I need to be more actively engaged and visible as a member of our community." He continues, "I still

IF I HAD A TIME MACHINE

"I would tell my younger self that I spend my days in front of a computer, and I have the world at my fingertips, just like I thought I did all those years ago with my encyclopedias. I get lost in the discovery, just like I did then, but now I have a laptop instead of a book, and I have the internet to give me access to more information than I could have imagined. I would also tell him that I get to talk to people all around the world who are interested in the same kinds of questions that I have, other scientists who are also trying to make sense and discover.

"When I was a kid the information in those encyclopedias about whales was very, very limited. It was mostly coming from whale hunting, which happened for centuries, until the whales were almost extinct. So I would also tell my younger self: You are going to be blown away! You'll be right in front of these animals, and they are alive and majestic, and so spectacular that it's going to be unlike anything you've ever seen. Because looking at the books and dreaming about them is cool, but actually getting to go and be in those places, and see those creatures . . . it's going to blow your mind."

felt that the lack of representation of LGBTQ folks in the sciences was indicative of [a problem]. Although not necessarily outwardly hostile, the professional societies that I belonged to did not provide any notion of being accepting of LGBTQ people."

So Daniel took action. In the article for *Discover Our Coast* he discusses how he came up with the idea of having an LGBTQ social mixer at a big scientific conference. Over 150 attended that first mixer, and it became a regular event. Daniel also seeks to play a supporting role on his own campus by mentoring and advising students from underrepresented groups, and providing a safe space for LGBTQ students to feel welcome.

"I wake up each day excited to work toward finding sustainable solutions for a green and equitable future."
—Devyani Singh

DEVYANI SINGH

Dr. Devyani Singh knows firsthand what climate change looks like. She grew up in the Indian Himalayas, the tallest mountains in the world, surrounded by nature and very much in the shadow of climate change. Increased glacial melt from the mountains, increased wildfires, and the near extinction of local animal species were just some of the climate challenges faced by her community. She grew up aware of the fragility of this special place, and entranced with nature and the environment. She says, "I used to get *National Geographic* magazine every month and read my aunt's encyclopedias, trying to learn everything I could about the environment and nature." Still, she says that based on images in media and culture, "a scientist was a person in a lab coat in the laboratory, holding a test tube. And I didn't want to do that. So even as I was interested in biology or ecology, I wouldn't have said I wanted to do science."

◀ Devyani Singh pictured with her dog, Zephyrus

THE HIMALAYAS AND CLIMATE CHANGE

The Himalayas are the tallest mountains in the world, home of Chomolungma, or Mount Everest as it's called in English, and dozens of other high peaks. They are a famous site for mountaineering feats, and a popular trekking destination for tourists from all over the world. But the Himalayas play a far more vital role than just tourism. They loom across eight countries, including Afghanistan, India, China, and Nepal, and well over a billion people depend on the water that flows from their glaciers into rivers like the Yangtze in China and the Ganga (Ganges) in India. This water irrigates crops, provides drinking water, and sustains life for people, animals, and plants.

These mountains are home to human settlements that have been there for thousands of years, as well as countless species of plants and animals, many of which are unique to this region. But they are all threatened by climate change. Rising temperatures mean increased wildfires and faster-melting ice, which can lead to floods, landslides, and other natural disasters. Organizations are working with local communities to help adapt to the changing climate. This might mean improving dams or building alternative waterways to handle flood runoff, planting new crops that are better suited to the changing conditions, or creating drinking-water reserves. The goal is to ensure that despite the changing climate, communities can continue to live among the mountains.

As Devyani got older, she still did not believe science was for her. "Growing up in my family in India, the options were to be an engineer, a doctor, or to go into business. Being a scientist wasn't really considered a choice. So I thought, I will study business, and make a lot of money, and then I can open a bunch of animal shelters! And that will be my way of doing the work I love."

Devyani followed this path, getting a degree in business and moving to the United States for her graduate program. She says, "I hated what I was doing. I was a senior financial analyst, working for a company that was shutting down factories in the United States and moving them to Mexico, in part because they could make more money by exploiting the lower environmental regulations there. I was making a lot of money at a very young age.

I was a success, but I was miserable. I was deeply depressed, because I was destroying the only thing I ever cared about . . . the Earth."

Devyani's sister helped her understand that her job was making her unhappy, and once she realized it, she says, "I didn't think twice. I quit. I had no idea what I would do next, so I took a road trip, driving across the United States." While driving in Yellowstone National Park, Devyani met a professor doing research on wolves. He was looking at the reintroduction of wolf populations into their original habitat, where they had been almost completely wiped out. They talked about his work, and Devyani told him how interested she was—especially because her dog in India was half wolf. The professor encouraged her interest, even though she explained she did not have a science background. She learned that there were many ways to approach science, including social sciences that study human dimensions of the environment. After that, she went back to do a second graduate program, in environmental science.

"That's when I reconnected with my love for the environment and nature, and understood that climate change was truly the crisis we needed to face. I realized that this is the work I always wanted to do: protect nature."

> **"I realized that this is the work I always wanted to do: protect nature."**

Going back to school was in some ways a continuation of the learning Devyani had always done on her own. "It was hard because I had never studied these subjects, but it was like reading my *National Geographic* magazines, or poring over the encyclopedias . . . I got to keep learning about all the things I had always cared about."

Despite feeling societal pressure to follow a certain path, Devyani said her family was supportive when she told them about her career change. She worried that her father would be disappointed; after all, she was quitting a successful career in business right as she was starting to rise. But instead he told her, "The planet needs more people like you. Do what you're passionate about, and we will help you."

As Devyani pursued her master's in environmental science, and then her PhD, she continued to do things differently than many scientists. While researchers often narrow their range of study to a very specific area, Devyani wasn't sure where she wanted to focus. Luckily her mentors helped her to understand that there were options. Her projects ranged from using her business skills on projects related to environmental finance, to working on sustainable forestry with First Nations in British Columbia, Canada, where she lives, to studying access to clean household energy in India. That project was particularly compelling for Devyani, who knows that in her home country a large percentage of the population still cook over open fires.

She continues, "I am not a very directed scientist, not an expert on any one thing. But these areas intersect, and I am able to make connections and draw conclusions from all the different areas I study. I let the universe lead!" She clarifies that it is important to have good mentors who are willing to let you follow your passions. Too often scientists can have their excitement deflated by supervisors who don't encourage them, so she urges students to seek out people who will support you and help you get the skills and tools you need.

WOOD FIRE COOKING ISN'T HEALTHY: FOR PEOPLE OR THE PLANET
Worldwide, over 50 percent of all wood harvested is used for cooking and heating, which has devastating environmental impacts. The smoke from these fires releases dangerous particulate matter that gets inhaled into the lungs and causes severe illness and over two million deaths per year, according the World Health Organization. And as women and girls are more likely to spend their time collecting wood and cooking, these issues affect them even more than the general population. So finding alternatives to cooking over wood is a critical environmental and health goal. A challenge Devyani embraces, "I wake up each day excited to work toward finding sustainable solutions for a green and equitable future."

Devyani is also clear that while she never wants to return to the business world, she is not always willing to stay within the traditional scientist role either. "I want to do more with science policy," she says, "helping ensure that politicians don't keep ignoring the science." Devyani herself has run for office from her home province of British Columbia as a member of the Green Party, a global political party that prioritizes the environment and the changing climate.

"I am definitely keeping a foot in politics," she says. "I can help get climate bills into the government or into policy, and work on getting them passed." She continues, "I struggle with the fact that there is a disconnect between academics and the real world. It can take years for traditional science to move from hypothesis to research to data analysis to publication. And we scientists are falling behind on how we communicate these important results! It is our duty to connect with the media and connect with policy makers! Sitting in our offices and doing science for the sake of science . . . that time has passed."

Devyani believes that while politics and climate science are inter-connected, there are times when they must also be kept separate. Specifically, she points out that when she does science, she collects data, assesses it, and then shares the results. But as a climate advocate, she goes further, draw-ing conclusions from the data and pointing out what it means for climate policy. She is careful to be clear when she is simply sharing facts and when she is offering her opinion and recommendations, because, as she says, "I don't want the science to be discredited or considered politicized."

When it comes to what she loves about her work, Devyani says that is easy. "I am surrounded by people working on such amazing research. And I am learning every day, learning about things I might not have known existed in the world. And then I get to collect data, and when we discover what the data are telling us, it's this wow moment! And then it's about finding real-world solutions to climate change." She jokes, "The things I love to do most, hiking and traveling—I would be doing them

"It is our duty to connect with the media and connect with policy makers! Sitting in our offices and doing science for the sake of science . . . that time has passed."

On the campaign trail. Devyani believes strongly that scientists need to get out of the lab and into the real world, because there is no time to waste in dealing with climate change.

When asked how she would describe her job to her younger self, Devyani says, "I would tell her I am a scientist who works with nature, and I'm trying to save animals and trees from future destruction due to our human activities, and trying to help humans who are most vulnerable to climate change."

anyway, but now I get paid to do it! I got to travel to over twenty countries for fieldwork and conferences. I got to travel to India and hike into the mountains that I love to research them. I don't make a lot of money. . . . I couldn't afford to travel like this on my own . But it's part of my job."

She says people probably think that as a climate scientist she's in a laboratory all day, running atmospheric models on computers, but in fact her work is often out in the field, and is very much based on human–climate interaction.

One thing that is important to Devyani is to be open about her whole self as a queer woman, as an immigrant, and as a scientist. She says, "In the business world I hid who I was, and was not open about my sexuality. When I moved into science I kept that part of me hidden, and I didn't tell anyone in my lab, or in my program. But slowly I told people, and they encouraged me to be open, and told me there would be support for me at the university. And they were right. Now I tell people in job interviews that I have a partner, and I speak out when I have a chance, because I think it's important for other people to see this." She continues, "The reason I am so vocal now, letting people know I am a queer immigrant scientist, is because when I was growing up I didn't have those kinds of role models to look up to."

"The biggest impact you can have is to make it known to politicians that you are paying attention. Write letters, protest, go to local council meetings and speak out about your future."

ADVICE FOR YOUNG PEOPLE

"The biggest impact you can have is to make it known to politicians that you are paying attention. Write letters, protest, go to local council meetings and speak out about your future. Groups like Sunrise Movement, Fridays for Future, and School Strike for Climate are making it clear that it's your future at stake.

"I wish everyone knew that it's not too late! There are doomsday folks who say that the planet is ruined and there's no point in taking action, but that's wrong. There is still hope, and we can still take action, and if we take action today, that's better than if we wait until tomorrow."

She says, "As climate scientists we talk about the different pathways that will get us to a different degree of planetary warming. There are different ways we can address the problems, and different choices that will impact what happens next. Optimism is all we have."

"There's a lot we can do to mitigate climate change, and the best thing I can do as a science communicator is spread the word: it's not helpful to be hopeless."
—Gabriela Serrato Marks

GABRIELA SERRATO MARKS

As a Mexican American, Gabi Serrato Marks always pronounces her first name *Gah-bee*, never *Gabby*. She earned her PhD from the Massachusetts Institute of Technology, which is one of the most prestigious science universities in the world. She has rappelled into caves and studied ancient stalagmites, but now she is a full-time science writer and disability advocate, using her knowledge and platform to help people better understand science, climate change, and the often unrecognized challenges people with disabilities face in the workforce.

Gabi believes her work as a science communicator is just as important as the work she was doing in the lab or in the field, since helping people truly understand the importance of the issues facing our world has never been

◀ Gabi pictured in a cave in Tamaulipas, Mexico, taking a water sample to study the chemicals naturally present in the water.

more vital. Still, making the switch from practicing scientist to writer was unexpected.

Gabi says, "As a kid I was always happy to be exploring, whether that was finding roly-poly bugs under rocks or riding my bike. But I also loved school, and thought about being a medical doctor. My aunt, who is much younger than my mom, graduated medical school when I was a kid, and I remember thinking that path might be for me."

It was in college, when she took her first oceanography class, that Gabi had a change of heart. "I realized in that moment that I could apply all the science that I enjoyed—all the physics and chemistry—with these environmental causes and places that I loved." She points out that she had positive, supportive role models, including an adviser who acted as a real mentor and "showed me what an academic career in geoscience or oceanography could look like."

"As a kid I was always happy to be exploring."

For her school science fair, Gabi dressed up like the famous Polish-French scientist Marie Curie.

Getting into MIT and pursuing her PhD was hard. Gabi says there were times she thought about leaving the program. "I had a conversation with my mom, telling her how hard it was and how I was thinking about quitting, and she basically said, 'You can consider quitting, but once you think of that option, it's always going to be in the back of your head. Sometimes you have to tell yourself that quitting isn't an option, and just keep going.' My mom is a lawyer who deals with incredibly challenging topics, like domestic violence and family law, so I know she's had her share of professional challenges. But I understood what she meant. I decided I would put my head down and finish my degree."

Gabi continues, "I have definitely dealt with imposter syndrome, where I feel I don't belong. Mostly it's feeling like I cannot perform at the level I'm supposed to. Being at MIT was a part of it, because the level of excellence is everywhere. So if you ever feel like you're *not* excellent, it can be really hard, because you feel like you're not living up to the expectations."

Gabi identities as part of the LGBTQ+ community, and while she did find some others in her program, she noticed that the majority of faculty, and certainly the most senior and most visible professors and provosts, were seemingly straight and in heterosexual relationships. "It's not like there were no others," Gabi says, "but I didn't see a lot of people like me."

Part of what was so challenging for Gabi were her ongoing and worsening health issues. She says, "My adviser at MIT tried to be supportive, as did my friends at school, but they didn't really know what I was going through." Instead, she says, "I found a lot of people online who have had similar experiences, and who were able to guide me through. They helped show me that there is a way to live really well and live a really happy life, even with chronic illness. The online disability community made a huge difference."

Gabi did not get an explanation for her health issues until she was in graduate school, when she was finally diagnosed correctly with Ehlers-Danlos syndrome, a genetic condition that affects connective tissue between bones

and can cause chronic pain and other health problems. She says it was hugely helpful to have that diagnosis. "It made such a difference in how I managed my condition and advocated for myself. Before I got the diagnosis I was a kid who had to go to the nurse a lot, who often got weird coughs or got sick, but I had no idea what it was."

She continues, "I think the disability community is honestly why I finished graduate school and stayed in science as long as I did. There's such an emphasis on advocating for yourself and others. The people who came before me advocated for things that didn't necessarily benefit them, whether it was automatic door openers or elevators that actually work. I followed those people, and worked to make sure I was advocating too. They also helped me realize that it's absolutely okay to ask for accommodation so I can perform at my best."

Gabi is clear that there is nothing wrong with needing to do things differently. She notes that for her, writing by hand can be challenging because of tendinitis in her wrists, so using a keyboard is much easier. When some professors didn't allow laptops in class, Gabi had to push to be allowed to type her notes

◄ Gabi rappels deeper into the cave with Jean Louis Lacaille Múzquiz, an expert caver, looking on for safety.

instead of write them. She says, "It was important for me to understand that not being able to write out my notes doesn't make me less good as a scientist, or even as a student. It's just a different way to do things. The disability community helped me frame my health issues more clearly, so that I understand that someone asking for a few extra days to read something, or access to speech-to-text software, is fine."

While it was challenging, Gabi did enjoy the work for her PhD. "I never thought I'd be someone who rappels into caves! I was a Girl Scout and loved it, but I was never a backpacker. And I'll be honest: I hate pooping outside! So being in the field for weeks was always a challenge. But I loved parts of it, especially because I did my fieldwork in Mexico, where my mom's family is from, so there was a sense of belonging."

"I never thought I'd be someone who rappels into caves!"

She says that some of the techniques she used to analyze rocks were the same observational tools that people have used for hundreds or even thousands of years, while others are more recent. One newer technique is uranium–thorium dating, where scientists look at how much of two specific elements—uranium and thorium—are present in a stalagmite, because uranium decays into thorium very slowly over time, so measuring quantities of each can give insights into how old the rocks are.

Some of the most exciting science happening now, Gabi says, has to do with the ability to take tiny microscale measurements of stalagmites and measure seasonal bands, like the rings of a tree. "The level of detail we're able to see is incredible," she says, "where we can see the effect of a wet season versus a dry season on how the rock formed. Before this new technology, we were lucky to see chunks that gave us data about decades, and now we can see seasons in one specific year. And this technology can be used for corals and stalagmites and other types of **paleo science**."

For someone who had been focused on a science career since college, and who had successfully pursued a degree at a prestigious university, Gabi's choice to leave the traditional science path of research and academia might seem unusual. For her, the idea of following that career trajectory, which often involves moving frequently for new job opportunities, did not appeal. Also, she says, "I felt that studying rocks and doing research in a lab was not the best way for me personally to make an impact. There are a lot of scientists who do make a real difference with their work, and we need them." But for Gabi, her real passion was writing.

Writing about science started almost by accident for Gabi. She never wrote for her college newspaper or had any experience, but she says, "I started writing about science and loved it. I loved talking to people for interviews. I loved the fast pace, turning a story around in a few months, unlike scientific papers, which often take years to write and publish. And I think it's so important to help more people understand

"I think I can make an impact by writing about science in a way that makes it more understandable and personal."

science, to understand why it matters so much." She continues, "I think I can make an impact by writing about science in a way that makes it more understandable and personal. For instance, as a Mexican American myself, it has been hard to hear all the recent US stereotypes and misinformation about Mexicans. So when I travel there I document it with photos and videos to give a more accurate picture. I am trying to shift the story."

The work Gabi does now still depends on her knowledge of science. And she says that when she's writing, she still uses the scientific method. "Just like in an experiment, when I'm writing a paper I figure out what my hypothesis is, then use it to frame my writing." In addition, she says, "I use many of the skills from my science degree—to organize huge amounts of facts and information, analyze data, and quickly focus on the most important elements when dealing with a large amount of information." Even in parts of

ADVICE FOR YOUNG PEOPLE

The most useful advice Gabi would give students is to keep track of their data. "I think this seems like a small thing, organizing your data, keeping good records, managing your notes and information, but it has come back to bite me several times! I have to remake graphs or research things that I already did earlier. So this is important in science and in life more generally—manage your information and keep track of it!" Also, she says, "If you have a chance to learn coding, definitely do. I learned in graduate school, but I wish someone had told me to take coding classes earlier. I would have been better off."

her job that she never expected—like managing social media accounts—she says her science training kicks in. "When I am creating posts for a client's Instagram or Twitter, it's the same kind of scientific experimentation. We try different techniques, tweak different variables, and look at the analytics to understand what's working, then iterate on that."

More broadly, Gabi says that she wishes she had known how challenging the path into science would be. "Sometimes the people who have already made it make it look easy. And when I looked at those role models I couldn't help worrying: What if it takes me an extra year to finish? What if I don't find a job? The truth is that there are often a lot of failures that we don't see that happen behind the scenes, or pivots that people make when their plans don't work out, but we don't know about them. So I would tell kids to be aware that it's not always as easy as some people make it look."

She continues, "I also would want kids to know: They are not alone. They are not the only one. Whether they are physically disabled or LGBTQ+, it can feel like you're the only person dealing with this stuff. But you're not. There is a whole wonderful community out there, full of people who are facing the challenges you're facing, who understand your experiences."

While there are people who do incredible things to break down barriers, Gabi says, "It's also okay not to be a trailblazer. You can just be yourself and do your best, and know that you deserve to be there as much as anyone else. And it's possible someday someone will look up to you and be glad to see someone who is just like them, studying the same kinds of science. Even if you don't feel like there are role models for you now . . . well, sometimes you have to be that person."

Gabi's job now isn't quite what she expected, but she loves what she does. She says, "I write about science and create videos and films and stories that help people get interested and informed about important things that are happening. And for me right now, that's more important than the paleoclimate research I was doing. Because there is such a need for science literacy! There's

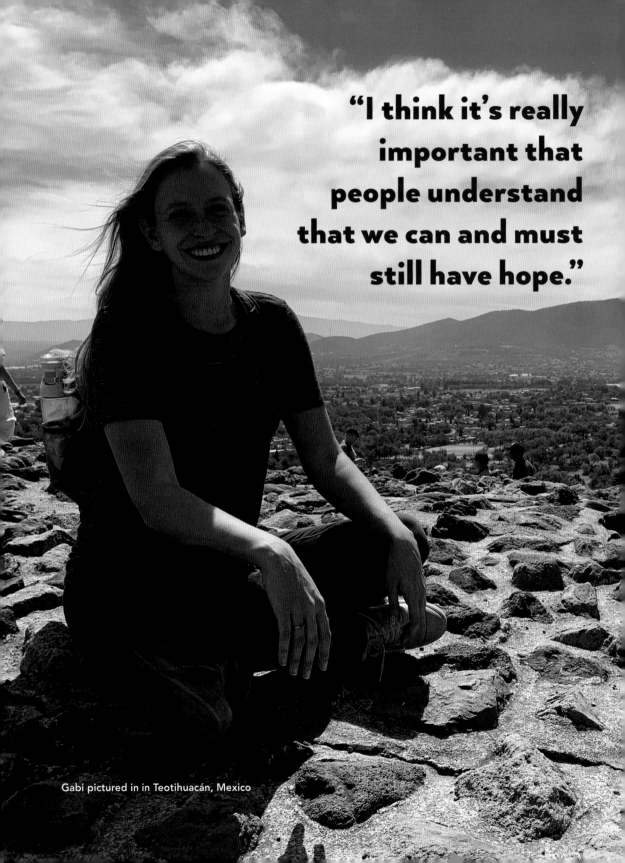

"I think it's really important that people understand that we can and must still have hope."

Gabi pictured in in Teotihuacán, Mexico

such a need for people to understand why it matters." She continues, "When writing about climate change and our planet, I try to avoid too much doom and gloom, because feeling hopeless doesn't help anything. We need to understand the problems, and there has to be change on a systemic level, but we can all also take personal steps to make a difference. It can be as small as conserving water, turning off the shower or faucet instead of letting it run. And then talking to family and friends to encourage them to change their habits. Because it's not one person's responsibility, but that doesn't mean we can't all take action."

Her main message for young people, whether they are interested in science or science writing or climate activism, is to remember the power of local work. "I think it's important for all of us to remember how much impact we can have in our own communities. Wherever you live, whoever is in your community . . . that's where you can make the most difference." She says, "While I am Mexican American and travel to Mexico every year, I am not a part of that culture or community the way people who live there are. And for that reason, I might not be the right person to communicate a message about climate change there." Instead, Gabi works to help people understand their power within their communities.

Gabi continues, "I think it's really important that people understand that we can and must still have hope. There's a lot we can do to mitigate climate change, and the best thing I can do as a science communicator is spread the word: it's not helpful to be hopeless. There are things we can do, and ways to be more resilient to the coming changes. But we have to be better about spreading that message."

Gabi isn't sure what her future holds. She says she might go back to practicing science, or keep writing, or something else entirely. She says, "I have no idea what my five- or ten-year plan looks like. And I think it's important to normalize that. . . . I still don't know what I want to be when I grow up! You don't have to have the whole plan figured out."

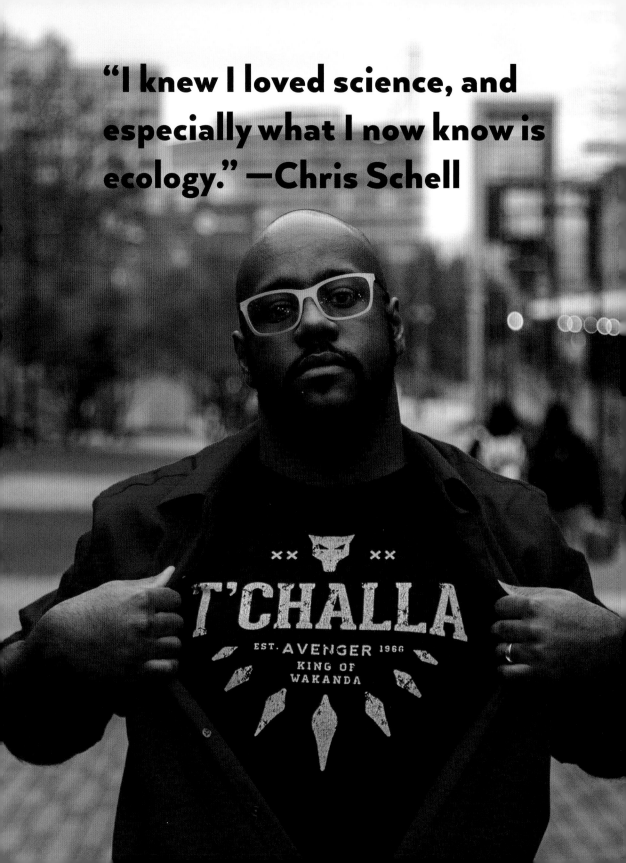

"I knew I loved science, and especially what I now know is ecology." —Chris Schell

CHRIS SCHELL

Chris Schell is an urban ecologist who studies coyotes and other wildlife that live in cities, and he pays attention to how humans and animals interact in urban settings. He's interested in all animals, and even thinks tarantulas are adorable. (He insists that if you zoom in on spider feet they look like little cat paws.) Originally from Los Angeles, California, Chris has done fieldwork in the Dominican Republic as well as Utah, Illinois, and Washington state, where he taught at the University of Tacoma.

Chris says he always loved science, and even as a kid "I was checking out slugs and snails and spiders and birds." He says his favorite shows as a kid were PBS animal shows, and that he loved dinosaurs—by the time he was seven he knew over a hundred dinosaurs by name and type. But for

◀ Bringing his whole self into science is important to Chris, and that includes his love of *Black Panther*.

Chris, who didn't know any ecologists or even any scientists that looked like him, his interest in science meant that he should be a doctor.

"I would compete in science fairs, and win first place, and I knew I loved science, and especially what I now know is ecology. But I didn't even know enough to call it that. There was a period of time, from the time I was twelve until I was in college, when I didn't have a connection to research scientists. Not to mention that very few folks in the environmental science space on TV or movies looked like me. So the members of my community would often suggest I would be a great medical doctor, and I internalized that, thinking that a field like neurosurgery was my passion."

Chris says he tried to follow a path to medicine. He went to Columbia University in New York, and volunteered at Saint Joseph's hospital, right near campus. But, he says, "I hated it. It wasn't exciting or stimulating, I think because I didn't have the freedom to be creative. And that's what I loved about those science fairs . . . the creativity to make an observation, come up with a hypothesis, build methods to test that hypothesis, and then analyze the results. It lets you really open up the gift of science as if it were Christmas morning—you don't know what you are going to get! Medicine didn't have that same appeal for me."

Chris acknowledges that for some doctors, including those who do medical research, there is that kind of discovery. But for him as a student, it was hard to see. And that's where the tarantulas come in.

"The summer after my freshman year of college there was an immersive ecology trip to the Dominican Republic. It included the chance to do an independent project, and that led to my first experience in field research. I had the opportunity to conduct an independent experiment, and chose to study spiders living around the research station where we were housed. Specifically, I compared the predatory behavior of golden silk orbweavers and Hispaniolan giant tarantulas in different environments. That was my first foray into the study of animal behavior in the field—and I loved it."

Now that Chris works with coyotes and not spiders it's easier to convince people that his subjects are cute.

When he returned to school in the fall, Chris was still on the premed path, but his heart wasn't in it. He struggled with the challenging classes, both because of how hard they were and because he wasn't interested. He says, "I remember calling my mom after a bad midterm and telling her that this just doesn't feel right. I told her how the work in the Dominican Republic felt effortless, but this . . . this wasn't me. I had tears in my eyes, because I was afraid I was going to disappoint my family. But they told me to do what I thought was best, and that they'd support me."

He acknowledges that he was lucky. "I've heard a lot from other colleagues of color that they've struggled with this cultural notion that science isn't for them. Not everyone got the warm and fuzzy reception from their families that I did. And honestly, it's true that people in my family still don't know what I do! Sometimes they'll say 'Oh, Chris studies wolves!' and I have to explain that no I don't, I study coyotes, which aren't the same thing. But that doesn't matter. They know I'm doing awesome and I love it."

When he switched from premed to ecology, Chris was lucky to connect with a wonderful mentor. He introduced himself after his first class, nervous but excited to talk about ecology and evolution. He says, "I would go to her office hours religiously, and was often the only one there. And I'd ask questions and talk about these subjects that I love." And while he says he sometimes felt behind in his classes, he didn't let that deter him. "I often tell

WHAT'S THE DIFFERENCE BETWEEN A WOLF AND A COYOTE?

Anyone who has ever watched cartoons has probably seen Wile E. Coyote chasing the Road Runner all through the cartoon desert. But real coyotes aren't cartoon bad guys, or wolves howling in remote forests. They are their own species, a fascinating one that lives in close proximity to humans.

Coyotes and wolves are distantly related members of the same animal family (Canidae) that includes the dogs we love as pets. But coyotes have a number of differences from their wilder wolf cousins. For instance:

- They are smaller.
- They have a higher-pitched howl.
- They eat both meat and plants.
- Their habitat is often much closer to human settlements.

This last one can be a problem for coyotes: as their habitats are bulldozed for new housing developments, shopping malls, or interstate highways, the animals have to adapt. Luckily coyotes are clever. They have learned to take advantage of human activity and find new food sources, from garbage to urban animals that are attracted to the area.

Another difference between wolves and coyotes is that wolves are far less likely to be seen by humans, especially in areas near cities and suburbs. Coyotes are increasingly common in human communities, often seen not only in local parks but in parking lots, streets, and backyards. While coyotes, like all wild animals, should be treated with caution, they are not a major threat to people. And given how quickly humans are encroaching on their land, it's important that we learn to live in community with coyotes. This means making sure that garbage cans are closed tight and small pets aren't left alone in the yard, so as not to invite coyotes in! It also means making noise on walking trails to discourage coyotes from getting too close. And of course, coyote pups, like dog puppies, are ridiculously adorable, but humans need to stay far away! Nobody wants to get up close and personal with an angry Mama Coyote.

"One thing I crack down on is the idea that you 'just have to want it enough.'"

my students that you might feel completely inferior, but if you really like it, keep at it. Just because you might do badly in one class doesn't mean that you can't pursue that subject, or that you don't belong there."

That first professor remained a mentor to Chris, helping him apply to graduate school and more generally find a path for himself as a professional scientist. He laughs. "Also, I went from tarantulas, which most people don't think are cute, to working with zebra finches in her lab, which are adorable, and then to coyotes, where I'm often handling the puppies. So the cuteness factor was huge."

Even with his mentor's support, Chris acknowledges that the path wasn't always easy. "There were times, especially in graduate school, when I was not at all sure I was on the right path. It was incredibly isolating, and during that time a friend and colleague died by suicide. I think it's important to understand how these environments can breed depression."

Chris stresses that unlike college, going to graduate school often comes with financial support that makes it possible to keep building a career in science. But at the same time, for students who come from low-income backgrounds, the amount of money may not be enough to keep them out of poverty. He says, "One thing I crack down on is the idea that you 'just have to want it enough.' That's playing into a handbook of **white supremacy**, because not everyone can afford to take a really low-paying graduate position." It's the job of the schools, not the students, Chris insists, to figure out how to ensure diverse folks from different backgrounds can survive and thrive while earning their degrees.

WHAT IS WHITE SUPREMACY?

White supremacy is a racist idea that suggests white people are superior to others of different races. It is a form of racism that has been seen throughout history, and still exists today. While extreme examples of white supremacy include slavery and segregation, there are countless more subtle ways that we are still fighting this battle. From school exams that are written in ways that make it more likely for white students to do well, to voter suppression laws that focus on limiting voting in Black communities, there are unfortunately far too many ways that political, educational, and legal systems benefit white society and discriminate against others.

Often these forces can be hard to recognize, and people don't even realize that they are benefiting from or being harmed by a white-supremacist system. But professors like Chris and many others are part of an ongoing effort to push back. By naming the problem and seeking to create systemic solutions on college campuses and beyond, people are working to dismantle white supremacy and create new systems that give everyone—regardless of race—an equal chance.

Graduate school in Chicago was particularly isolating, Chris says, because he was one of two Black students in the whole program. He says the lessons he learned there help him be a better mentor to students now. "I realize now I had a level of depression that was undiagnosed while I was a graduate student, and some of that stemmed from my inexperience in the biological sciences space. I didn't know the process of publishing a paper . . . or building community relationships within academia. . . . My other Black colleague and I felt we were always playing catch-up to an unwritten standard. The way we did science was undervalued. I know

> "My other Black colleague and I felt we were always playing catch-up to an unwritten standard. The way we did science was undervalued."

that for my friends and colleagues who are Black women in academia, the challenges are even greater. They experience numerous microaggressions, and are often treated as though they don't deserve to be in their programs."

He continues, "I was shielded a little bit because I was working with a professor who became one of my best friends and mentors. She showed me what it looks like to break down barriers for marginalized folks. . . . She would see when I was stuck, then step in." He credits her for helping him create an experimental design—basically designing a research experiment as carefully as possible to ensure the data would answer the intended question—when he was struggling. Chris says, "She would help me drill down and ask the right questions."

He also says that the work itself was exciting. "Graduate school is nothing like undergrad. You get to build and create, and that freedom is magical. You have the flexibility to say 'I really like this question, and I want to pursue it' and then go figure it out. I loved that freedom, though there are other folks who prefer more structure, and more direction."

Chris and his one Black colleague worked to create a more welcoming space at their university, ultimately forming the Graduate Diversity Committee, which at the time was unique. But Chris says that their efforts weren't supported by the more senior faculty, most of whom were white men. He says, "They believed that science is science and should be objective, so there was no need to talk about diversity. But they're wrong!" Chris and other scientists point out that personal experience always influences scientific study. The problem is that many white male scientists have been told their whole life that their point of view is the objective norm, not their subjective experience. This leads them to dismiss attempts to talk about the importance of diversity.

Chris says that often when researchers are studying wildlife interactions in urban environments, they focus on specific types of neighborhoods and communities. Not surprisingly, white researchers typically study white areas of the city, paying less attention to communities of color. The result is that they draw conclusions about where wildlife is—and isn't—based on incomplete data. So they might conclude that coyotes are only in certain areas, which is untrue. It is simply a factor of where the scientists chose to do their research. This is an example of the kind of hidden racial bias that occurs often in science.

Now, as a professor of undergraduate students, Chris works hard to encourage people to bring their whole selves into science. He says, "Incorporating diversity, equity, and inclusion into every part of the university, from the labs to the adviser relationships to the student support, doesn't just improve morale. It improves the science. It encourages people to be open to doing science in ways it's never been done before, and that makes science better, and makes scientists better."

As he does with other elements of science, Chris pushes back on the idea that all writing has to sound the same to be scientific. "You have to learn the rules, you have to know how to present yourself, but then we should be allowed to portray ourselves honestly. Your style is your own, and that should be allowed."

What Chris finds exciting is that, since he has started talking about the importance of diverse perspectives in science, he has found a large and welcoming community of all kinds of scientists who are eager to have this conversation. "These are scientists who might be studying slugs, or predator–prey interactions in salamanders, and suddenly we're talking about how to dismantle white supremacy and promote Black excellence and ensure we're supporting our transgender and other LGBTQ+ colleagues. I am so glad that we're finally ignoring the folks who said we should only pay attention to the science, because it is objective. It's just not true. And these conversations are important."

Chris relates his work in ecology back to his work as an advocate for more diversity in science, saying, "If we want better scientists, and if we want a better understanding of the natural world, we need to speak to the community members that have traditional ecological knowledge." He continues, "I want kids to know that their experience is valid, and welcome, and needed. As an example, when I was a kid, growing up in Los Angeles, I knew that climate change was a problem. There were already forest fires, and it was hot, and I knew this was going to get worse, but my voice wasn't recognized as being a part of the climate change conversation. Now we're expanding that conversation to talk about climate justice, and to talk about how communities of color, that have maybe experienced rolling blackouts, or water shortages, or are part of urban heat islands, how they're going to have it worse. So these experiences, my voice, it matters."

He continues, "I want students of color, and especially Black students, to know that their lived experiences are valuable and are going to help save the planet someday. The evidence is building that shows how systemic racism

"I want students of color, and especially Black students, to know that their lived experiences are valuable and are going to help save the planet someday."

WHAT IS BIODIVERSITY?

When following issues of climate change we often come across the terms **biodiversity** or biodiversity loss. But what exactly are we talking about? Basically the word is a combination of two important terms: "biological," which refers to everything that has life on Earth, and "diversity," which, in science, relates to the variations among different living things. So biodiversity is nothing less than the vast variety of life on our planet, from invisibly small microorganisms to animals and plants and insects and mushrooms and spiders and . . . you get the idea. EVERYTHING.

So why does it matter? There are lots of reasons, some more obvious than others. On the most obvious level, humans live as part of a big interconnected web of life, and our existence and comfort are related in large part to other species. For instance, I don't love getting bee stings (stepped on one at summer camp when I was eight), but bees play an enormously important role in pollinating the plants we eat. Pesticides that kill bees mean fewer bees to pollinate apple orchards, strawberry fields, corn fields, and so on. Which means farmers have a harder time growing food. Which means they struggle to make enough money to keep their farms going. Which means prices for food go up, and families like yours and mine have to pay more at the grocery store. Which might make it harder to pay for other things, like soccer cleats and books. All because of the bees.

And that's just one example. But the importance of biodiversity goes beyond the impact on humans. The reason we have to stop species from disappearing is not just because I don't want to pay more for my potato chips or strawberry ice cream! Humans might be controlling the planet but we are one of millions of species who call it home, and we are not the only ones who matter. When a songbird species goes extinct, or when a certain fish disappears forever from the oceans where it swam, the whole world suffers.

As humans overrun more and more of the planet, we dramatically change the landscapes we inhabit. We cut down trees and build dams that disrupt rivers. We build tall skyscrapers that change how birds migrate and we pave over open spaces. While there are often good reasons for our actions, and humans have created some amazing things, there's no question that human activity has eaten up the natural world, and with that disappearance come huge threats to millions of species. The solution is not simply to get rid of the humans, or even to have less humans on the planet, but instead to figure out how to live equitably with all other living things

So what can we do to prevent biodiversity loss? First of all, it's important not to feel guilty just for existing! Yes, humans are a big part of the problem, but we can be part of the solution too. We can help conserve land near our homes, plant bee- and butterfly-friendly wildflowers in a vacant lot, create a community garden, do a beach cleanup, or talk to our families about conserving water and electricity. All of these small steps add up to changes in a community, in a town, in a state, in a country . . . and eventually, around the world.

and oppression actually generates changes in the environment that lead to biodiversity loss in a profound way. So in order for us to be better stewards of nature, in order for us to be better conservationists, in order for us to be better biologists or scientists, we have to solve our social ills."

The negative environmental effects of pollution, climate change, and urban sprawl are felt in oppressed communities, but also in nature. Research done by Chris and his colleagues has found data to substantiate the idea that species other than humans are negatively impacted by white supremacist laws and policies. The same systemic problems that lead to inequality in housing, food accessibility, damage from climate change, and other human concerns also have a negative impact on biodiversity. For example, systems of racism and classism lead certain neighborhoods to have fewer trees, less green space like parks, and more environmental con- taminants. That in turn directly impacts the wildlife species—like coyotes— that live in urban environments. According to Chris, studying these species without acknowledging social justice issues leads to incomplete science.

As Chris noted, science itself isn't objective, and when most studies are done in majority-white, wealthier areas, scientists are missing key informa- tion about the effects of climate change and population loss in other areas. Chris says, in an interview for *Berkeley News* about his recent research paper that highlights these inequities, "I hope that this paper will shine the light and create a paradigm shift in science. That means fundamentally changing how researchers do their science, which questions they ask, and realizing that their usual set of questions might be incomplete." He says that doing this means, "We will not just have more diversity in science, we will have better science and better scientists."

"I grew up with the idea that science was just the coolest and most fun thing anyone could possibly study."
—Katharine Hayhoe

KATHARINE HAYHOE

Katharine is an **atmospheric scientist** who studies climate change and why it matters to humans. She's Canadian, but now lives and works in Texas. She has earned many prestigious awards for her science, including being named one of *Time* magazine's 100 Most Influential People. Katharine is famous for her clear and easy-to-understand discussions of climate change, and is often on TV or being quoted in the newspaper. She also gets tons of hate mail—on bad days, more than two hundred pieces a day—from people who get angry because she says that climate change is real and caused by human behavior.

◀ In Texas, where Katharine lives, talking about clean energy like wind turbines isn't always popular, but she pushes past the assumptions to what people really care about: their land, their families, their communities.

WHAT IS ATMOSPHERIC SCIENCE?

Atmospheric science includes several different kinds of science, so it is sometimes called an interdisciplinary field of study. It uses the tools of chemistry and physics to study elements of the Earth's atmosphere. Some fields that atmospheric scientists typically focus on include meteorology, the study and forecasting of weather; climatology, the study of more long-term patterns in Earth's atmosphere; and aerometry, which focuses more on the upper atmosphere. These areas overlap, especially around the issue of how the changing climate is impacting our communities. Katharine received her master's in the area of atmospheric chemistry, and her PhD focused on climate dynamics.

Katharine says she doesn't often feel famous. When she received a *Time* magazine award, everyone in her family had the stomach flu, and she was cleaning up vomit when they called and told her the news. Four weeks later she was on a red carpet with photographers snapping photos, wearing a dress that was so long she had to buy extra-high heels to avoid tripping over it. She says that's just one of the weird things that happens when you're a mom, a professor, a famous climate scientist, and an evangelical Christian. Often the different elements of her life can seem to be contradictory, but Katharine wouldn't want it any other way.

When she went to university, Katharine knew she wanted to study science, and was interested in space. She says, "I was particularly interested in discovery, and for that, astronomy is just where you want to be. With the advances in instruments, we are finding new things all the time . . . it's like a buffet of discovery! But to finish out my degree I needed one more elective, and I decided to take a class on climate science. And it changed the trajectory of my whole life."

She continues, "It wasn't that climate science offered more opportunity for discovery . . . but that love of discovery was more for me, rather than any real desire to benefit humanity.

SCHOOL DAYS

Katharine always loved science, even as a kid. She says, "My dad is a science teacher and my grandmother had a degree in science education, so I had an appreciation of science my whole life. My dad is one of those people who is absolutely passionate about science—if he sees something he doesn't understand, he will stop and spend as long as he needs to figure it out."

She remembers as a child going to the park at night—"It felt like the middle of the night, but was probably 9:30 or 10:00 p.m."—and lying on a blanket looking at the sky while her dad showed her how to find the Andromeda Galaxy with binoculars. "So I grew up with the idea that science was just the coolest and most fun thing anyone could possibly study. It wasn't until I got to high school that I started to hear from my friends 'Oh, that class is too hard,' 'Girls don't take physics,' and stuff like that." She laughs. "And in fairness, some of those teachers made the subjects about as dry and boring as possible. Like physics! That's a class that's absolutely full of experiments and curiosity and wonder, but the teacher . . . well, he had no passion. He was just reproducing information."

"I had an appreciation of science my whole life."

Katharine grew up with science teachers in her family, and they often explored the natural world together.

And what I learned in that climate class is that climate change is the great threat multiplier! It's taking everything we're worried about with poverty, disease, lack of access to clean water, biodiversity loss, extinction, the future of human civilization . . . and it's making them worse. And the skills used to study climate change were the exact same skills that I had been learning in physics and astronomy. So I felt like, how can I *not* do this work, given that I have these abilities? How can I not do everything I can to help fix this enormous global problem?"

Katharine acknowledges that it was a move away from the excitement of discovery that astronomy promised. "My area of climate science is more like the medical field, where you're not trying to find a cure for cancer, you're trying to use all the tools that exist and make sure they're accessible and used correctly to help people take care of a problem before it's too late. Someone eventually might find a cure, but meanwhile we have a lot that can be done with the tools we have." She figured that she could help humanity use the tools we have to fight back against the problems climate change was causing.

When Katharine was in her undergraduate program and learning about these problems, such as the fact that climate change was pushing more people into poverty and causing suffering around the world, it seemed obvious to her that society needed to take quick action. She thought, "This is such an urgent problem! Surely we will fix it soon and then I can get back to astrophysics." Now, almost thirty years later, that seems like wishful thinking.

After switching her area of study, Katharine knew she would be heading to graduate school to learn more and figure out how she could help. She says, "I realized quickly that I didn't just want to study climate science, researching and writing academic papers. I wanted to work on the policy side and make a difference. When you go to graduate school your advisers direct your research path and really guide your career. I knew I needed someone who was not only a good scientist, but who understood that the reason we're studying this is because we want to fix the problem in real life."

When students apply for graduate school they often visit various schools and meet the different professors they could work with. At one school she interviewed with a professor who had worked with governments, nonprofits, and even big corporations on the use of chemicals and their impact on the environment. As soon as she met him, Katharine knew this was the place for her. He became her graduate adviser, and they still work together. Katharine says that, for her, finding someone who shared her work interests—doing science that makes a difference—was critical.

While Katharine's journey into science was fairly straightforward, she says there was a time she thought seriously about quitting. It was after she read a book called *Climate Cover-Up*, all about how polluting companies used public-relations strategies and advertising techniques to lie to the public about the dangers of climate change. This was not new; corporations had been using these tools for years, discrediting scientists, muddying the water around the data of climate change, and discouraging climate action. But after realizing just how smart and how extensive and how well-funded this misinformation strategy was, Katharine says, "I wanted to give up. I felt like I had no business trying to fight back—that I didn't have the tools or the expertise or knowledge to combat this multimillion-dollar disinformation campaign. I felt like a Girl Guide [a Canadian Girl Scout] trying to go up against Army Special Ops."

At the same time she was reading this book, Katharine was dealing with one of her earliest experiences of being discredited and publicly attacked for

"I wanted to give up. I felt like I had no business trying to fight back."

"I will *not* let these politicians beat me. It was then that I decided, if they're going to use this playbook, I'm going to find it. And I'm going to read it, and use it, and beat them at their own game."

her view on climate change. She had been asked to write a chapter on the importance of climate action for a book written by a prominent Republican congressman. She wrote the chapter, but while the book was being edited, the congressman decided to run for president. As he ran his campaign, he turned his back on climate science and very publicly rejected Katharine's work and her inclusion in the book. Suddenly conservative radio hosts were insulting her on their shows, and hate mail came pouring in. She wondered, *Why am I doing this? What's the point?*

WHAT IS SOCIAL SCIENCE?
Unlike physical or life sciences like astronomy or ecology, **social sciences** focus more on human behavior, and the relationships among humans and the institutions that make up our societies. These include how we interact with each other in government, the economy, family life, and more. Some common topics that social scientists study include political science, sociology, economics, and anthropology. Like other scientists, social scientists ask questions and collect data to come up with answers. But the questions are more focused on people and how we live together, how we make decisions, how we form opinions or biases, and so on. So for instance, medical scientists might determine that smoking causes lung cancer, but social scientists might try to understand why people continue to smoke if they know it will cause a disease that will kill them.

But then, she says, "I thought, I will *not* let these politicians beat me." Katharine continues, "It was then that I decided, if they're going to use this playbook, I'm going to find it. And I'm going to read it, and use it, and beat them at their own game. That's when I made a very conscious decision to pivot my area of study, again. This time I wanted to dive into the **social sciences**, looking more closely at how information gets shared, how political discourse happens, engaging with social scientists that study things like how we take in information."

Katharine started learning more about public relations and messaging, as well as engaging with researchers in sociology and political science. Using her same scientific lens, she worked on framing her message about climate change, then studied the results to see which messages seemed to reach people best.

Katharine acknowledges that the tools of communication and marketing are not inherently good or bad. You can use them to sell cigarettes, or to convince people to wear their seat belts. The bad part is when they are used to run a multimillion-dollar campaign to convince people that the scientists are wrong or that facts don't matter. That's why she decided to start using those tools herself.

Still, even as she has successfully worked to change her messaging to reach the most people possible, Katharine still gets hate mail every day. She says, "I think it's important to recognize the importance of this work, especially in the face of anger and pushback. Other people's anger is not an accurate assessment of the value of my work. There are many more who appreciate what I'm doing, even if they're not as loud. I've had to remember that."

Since that dark moment when she thought about quitting, Katharine's life has changed in dramatic ways. She explains that she divides her work into three buckets. "The first part is doing the science, which means experiments, analyzing data, writing papers, writing grants, and working with both graduate and undergraduate students where I teach. I also present at scientific conferences and serve on advisory panels for science organizations.

FROM THE 1600s TO NOW: TOOLS OF CLIMATE SCIENTISTS

One of the mistaken beliefs Katharine says exists about scientists is that they are all in the laboratory, wearing white lab coats and pouring things from one beaker to another. She jokes that she doesn't even own a white lab coat, and that all her work is in the field or with her computer, not in a lab. One interesting thing about her work is that it always involves a mix of old and new techniques and tools. For instance, a large part of her work is using data measured by thermometers, which have been around since the 1600s!

But another part of her work involves supercomputers, which are machines capable of taking huge amounts of data—think about twenty million filing cabinets full of information!—and analyzing it to predict what our climate might do, depending on the choices we make. These computers are so powerful that they run programs that would have been impossible even ten years ago. But the technology has improved, and now scientists use these climate models to predict what the climate might do: how much rain might increase over previous years, or how hard the hurricane winds might blow. Katharine and her colleagues use these supercomputers to model different scenarios of human behavior and predict the potential damage that might be caused as a result.

In 2021 two early climate modelers, Syukuro Manabe and Klaus Hasselmann, won the Nobel Prize in physics for the work they did in this field in the 1950s and 1960s. Katharine says being able to model climate is still yielding exciting new insights in the fight against climate change. For example, climate models are essential tools in a field known as "attribution of extremes," which is a fancy way of saying "coming up with a way to put a number on how climate change made a specific weather event worse."

Scientists have already figured out that just ninety corporations are responsible for two-thirds of carbon emissions since the beginning of the industrial era. If we know that climate change caused billions of dollars of damage by making a hurricane worse, and we know that ninety companies are mostly responsible for the carbon emissions that caused that climate change, then the companies responsible might start to get stuck with the bill for fixing the damage. And once the problems of climate change start costing them money, corporations might work harder to improve their business practices.

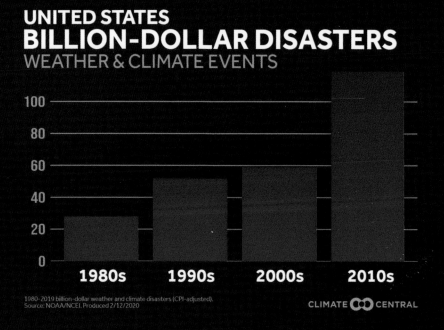

UNITED STATES
BILLION-DOLLAR DISASTERS
WEATHER & CLIMATE EVENTS

100
80
60
40
20
0

1980s   1990s   2000s   2010s

1980-2019 billion-dollar weather and climate disasters (CPI-adjusted).
Source: NOAA/NCEI. Produced 2/12/2020

CLIMATE ⬤⬤ CENTRAL

This chart, based on data from the National Oceanic and Atmospheric Administration, shows the cost of weather- and climate-related disasters in billions of dollars.

"The second part of my job," Katharine explains, "is being a scientist in the broader world. The subject I study is relevant to so many businesses and governments, and impacts decisions they make. I help them incorporate climate-change information into their planning, whether that's snow-making for a ski resort or drought conditions and water allocation for agriculture. I've analyzed flood zones for airports and city planners, and talked crop techniques with farmers. These are real-world decisions, and I help them find the science they need to make smart decisions for the future." In 2021 she became the chief scientist for the Nature Conservancy, the largest conservation organization in the world. It works in over seventy countries around the world, and Katharine is responsible for its global climate advocacy and adaptation work.

And finally there is the third part, which began more than ten years after she read that book. "The last third of my work is outreach and engagement. This means I'm often on social media, or giving a TED Talk, or writing an essay for *Time* magazine. I have a PBS digital show on YouTube called *Global Weirding*. In 2021 I wrote a book called *Saving Us* that explains how everyone has a role to play in fixing climate change. I helped launch a new campaign called Science Moms that connects caring about your kids and their future to climate change. And maybe most importantly, I give a lot of talks, over one hundred a year, to everyone from church groups to businesses, from rural farm groups to schools."

At a talk she gave, Katharine asked the audience, "How do you feel about climate change in just one word?" The image on top shows their responses at the beginning of the talk; the image on the bottom, after the talk. In this instance, talking about climate change is the first step toward taking action.

"The conflict with climate change arises because many who are Christian in the US have confused their faith with their politics."

One of the things Katharine is most famous for is that in many of her talks she is able to connect religious communities with the science of climate change. As an evangelical Christian married to a pastor, Katharine recognizes that, in America at least, her beliefs might seem contradictory. But she is quick to push back against this idea. "I grew up with the idea that the Bible is God's written word, and nature is God's created world. Christian faith, which calls for us to tend the garden and be good stewards of the gifts that God has given us, is not in conflict with climate science."

Katharine points out that she is not alone in combining science with her religious faith. "A study of researchers at top US universities found that 70 percent of scientists are spiritual people and 50 percent identify with a specific religious tradition, whether that's Christianity, Judaism, Islam, Hinduism, Buddhism, or another faith. The conflict with climate change arises because many who are Christian in the US have confused their faith with their politics."

So how does Katharine use what she's learned about messaging, framing, and communication to share her message with reluctant and sometimes hostile Christian audiences? She laughs. "Well, first of all, you never make any friends by telling people they're wrong. I try not to approach

"You never make any friends by telling people they're wrong. I try not to approach people with judgment. Instead I have to find what we have in common, and what we both care about."

people with judgment. Instead I have to find what we have in common, and what we both care about. I don't want to change people—I want to show them that this matters because of what they already care about . . . like their families, their land, their way of life."

She continues, "Science and faith can go hand in hand. Think of science like a compass. It can point out which way is north, but which is the right path to get there? We need a map for that. So science can tell us that the planet is warming and human activity is responsible. But what are we going to do about it?"

For Katharine, that's where her religion and her faith come in. She believes that many religions, not just Christianity, value caring for the poor and vulnerable and caring for nature, and that these values can provide a frame to help make decisions based on the science.

Katharine often talks with students raised in conservative Christian families in the United States, who then went away to college and learned that what they had been told about climate change was false. She offers several resources for students looking to find like-minded community, from BioLogos, which offers a scientifically sound Christian homeschool curriculum, to groups like Young Evangelicals for Climate Action and the American Scientific Affiliation. She says, "I hear heartbreaking stories of kids who grew up being taught false information about science, and then left home and learned differently. They feel so betrayed. . . . I think it is really important for kids and young adults to know that if you are Christian, and you're interested in science, and in climate change, there is no incompatibility of any kind, no matter what anyone tells you! If you were raised as a Christian and you want to pursue science, you are doing the exact right thing. And you're not alone."

ADVICE FOR YOUNG PEOPLE
The one thing Katharine wishes she had realized sooner is that "you don't have to have your whole life mapped out. Everyone talks about five-year plans and ten-year goals, and life goals, and I'm doing just fine with none of these. All you need to worry about is the next step; just the next one, nothing more. Make sure you're headed in the right direction for that time in your life, and let the future take care of itself."

"Make sure you're headed in the right direction for that time in your life, and let the future take care of itself."

"I had always dreamed of getting my PhD, even as a teen, but I was told that path wasn't for me." —Valerie Small

VALERIE SMALL

Valerie Small's path into science was anything but straight and narrow. She now has a doctorate from Colorado State University and is an expert on both the cultural and ecological importance of native trees and plants on tribal lands, as well as tools that help predict how these species will react to the changing climate. Valerie is a an enrolled descendant of the Crow Tribe (Apsáalooke) and loves to bring together scientific and Indigenous learning.

For Valerie, an interest in science started in her backyard when she was a kid, and now she says that part of her work is to ensure there are plains cottonwoods and traditional food species such as chokecherry, buffalo berry, and wild plums to improve ecosystem diversity along riverbanks

◀ From the time she was a teen to her time now as a grandmother, Valerie kept working to stay on her path into science.

"When I was little, I didn't call it science. I was just immersed every day within the natural world."

and provide traditional food sources for generations yet to come. Between her childhood and her grandchildren's, Valerie has studied many years to pursue her dream of being a scientist.

She says, "When I was little, I didn't call it science. I was just immersed every day within the natural world. I grew up spending time watching insects like the cicadas that would leave their shells hanging behind on trees, or watching the tiny fish in the creek that I played in with my cousins. My family grew large gardens and my mother canned food, all of which I found fascinating to watch and learn."

While Valerie loved biology in high school, she says that most women in her community who worked were secretaries or assistants—otherwise they were considered extraordinary. She says, "I was an honors student, but was told I would never be an A student in science, and I didn't test well on my standardized tests, so I was not encouraged to go to college." In addition, Valerie became a mother while she was still a teen, and, as she says, "After that it was about surviving and figuring out what came next."

Valerie started working in a science-adjacent field, health administration. She says, "My mom became a nurse when I was in my early twenties, and I really admired her. My father never had much formal education. He grew up on a farm, then was in the Navy, and when he retired, he took classes to learn motor mechanics. I had these role models of people finding their way and continuing to learn. I had always dreamed of getting my PhD, even as a teen, but I was told that path wasn't for me."

Still, with her parents modeling that career change and further education were possible, and with her own determination to pursue her scientific curiosity, Valerie got there eventually.

When Valerie was thirty-two, she went back to school. "I took one anthropology class, and at the end of the first class I walked out and changed my major." Valerie switched her major to anthropology and studied feeding behavior and parasite loads during her fieldwork with mantled howler monkeys in Costa Rica. But once again, Valerie's path did not go as planned. Her third child was born with autism, a developmental disability, and she knew she would need to study something closer to home. She rediscovered her love of plants and trees, and chose plant ecology for her master of science degree.

While Valerie grew up moving around the country for her father's Navy career, she knew her mother had spent her early years on the Crow (Apsáalooke) Indian Reservation. Valerie kept in touch with family members and eventually returned to her mother's side of the family in Montana, wanting to give back to her tribe and learn about the land and culture where her family was from.

Ultimately Valerie studied the ecology of the land while working for the tribal college in Crow Agency, Montana. She sought to give voice to the Crow people, integrating science and the traditional knowledge. She studied the

She sought to give voice to the Crow people, integrating science and the traditional knowledge.

culturally significant cottonwood trees, sacred to the tribe and used in rituals and particularly important for the Sun Dance ceremony. She studied how these trees were being threatened by Russian olive trees, an **invasive species** planted by settlers. She says, "Looking at the way the climate is changing, I utilized a habitat prediction modeling tool using climate variables to determine future spread of this invasive tree species. There is a real risk that the cottonwood trees will no longer be available to harvest for ceremony, given the spread of this invasive species and reduced water flows due to climate change. They are the canaries in the coal mine for this sacred land, and they are warning us that we could lose the trees and plants that we use for our traditional foods, medicines, and ceremony."

Cottonwood trees are a species of trees native to North America. They are sacred to the Apsáalooke or Crow tribe.

The silvery-green Russian olive tree pictured along the Yellowstone River in Montana, an area native to cottonwood trees

WHAT ARE INVASIVE SPECIES?

Indigenous plants are plants that occur naturally in an environment and were not introduced or planted by humans. Like Indigenous communities, these plants have been displaced and are often at risk. Sometimes the displacement is on purpose, as when a unique wetland is filled in so that buildings can be erected, and sometimes it is by accident, like when **invasive species** thrive and kill off the indigenous plants by taking all the resources like soil, space, and water. Invasive species can be plants, animals, or even tiny organisms. They are defined as something that is not native to an ecosystem but was introduced there—and, just as importantly, their introduction causes harm to the environment. Certain plants are invasive species, spreading quickly and killing off native plants. An insect can be an invasive species that spreads disease to humans or other animals. A fish can be an invasive species that takes over an environment and threatens the food supply for native fish. Because they are newcomers, these species often don't have any natural predators, and therefore can spread unchecked.

Invasive species can spread without anyone realizing, like when someone brings firewood with small insects into a new part of the country and the insects make themselves at home. Or they can spread through thoughtlessness, like when people release pet pythons in the wild when they grow too large. (This really happens! Florida has a major python problem!) Most invasive species are introduced by human activity.

"I thought, oh no. I am an Indigenous woman, and I'm older than most students, and although it might be more challenging to find a job, I am doing this. This was my dream since I was a teenager."

It was incredibly challenging for Valerie to care for her growing family, which included her stepchildren, and keep working toward her PhD. At one point a colleague, watching her struggle, said, "You know, Val, maybe a PhD isn't for you." The effect was the opposite of what he intended.

"I thought, oh no. I am an Indigenous woman, and I'm older than most students, and although it might be more challenging to find a job, I am doing this. This was my dream since I was a teenager."

Unlike many scientists, when Valerie completed her dissertation and got her PhD, she did not follow the conventional track to academia and teaching. Instead she wanted to look at land policies as they relate to tribal lands, and how climate change and agriculture affect native plants. She wants to improve access and availability for tribal members to be able to harvest plant species that are culturally important, so that traditional foods, rituals, and landscapes can continue.

For Valerie's work with the tribes in her region, community engagement and education is key. Part of her work is to talk with people about the future: the use of renewable energy resources and what a more sustainable future for the planet might require of us. She says, "My work is tribal developed and led, as each tribe has their own set of cultural priorities given their unique histories with the land." For instance, in one project Valerie and her team connected the community through a solar education program and invited community Elders to speak at these workshops.

Valerie says her work is a mix of cutting-edge science led by traditional knowledge. She collects knowledge from community Elders to learn what the land was like long ago, then uses tools like **computer algorithms**, **predictive modeling**, **habitat prediction**, and **climate models** to combine this knowledge and inform tribal land planning so that it is adaptive and can shift in response to extreme weather events and the impacts of ever-increasing global warming.

Bringing together the larger community and engaging them in conversation was a big part of Valerie's work with a nonprofit organization called Trees, Water & People (TWP). Tree planting projects developed and led by the tribes bring communities together to increase food security and ensure continued availability of culturally significant plants and animals.

"It's about having youth and Elders come together to foster intergenerational transfer of traditional knowledge—language, culture, and place."

"It's about having youth and Elders come together to foster intergenerational transfer of traditional knowledge—language, culture, and place."

PREDICTIVE HABITAT COMPUTER MODELING ALGORITHMS . . . WAIT, WHAT?

When you first come across scientific language it can feel like you've fallen into a bowl of alphabet soup. Science jargon, as it's sometimes called, can be confusing and cause people to tune out, because they think it's too complicated. But often big words have fairly simple meanings (after all, you can say "It traveled directly to my cerebellum" or "It went right to my head!" and it means the same thing). So let's take them one at a time.

Computer algorithms are, at their most basic, a set of instructions for how a computer does a job. For humans, an algorithm is a list telling you what to do. Say your job is doing the laundry. There are a bunch of different instructions you can follow to do it: you can bring clothes to the laundromat, you can put them in a washing machine at home, you can wash them in a tub of soap and water . . . you get the idea. Just having the job doesn't tell you how to do it! It's the same for a computer. If scientists need a job done, like sorting millions of numbers or analyzing years' worth of data, powerful computers can do in minutes what would take humans weeks or months! But how the computer does the job can matter a lot, so using an algorithm that will get the best, most useful result is important. Computer programs are mostly a series of algorithms that give detailed directions on how to complete each task.

Predictive modeling is a technique used by scientists to predict future behavior. It works by gathering large amounts of information, or data, and feeding it into a powerful computer that analyzes historical patterns and current updates to generate a model of what is possible in the future. For climate science, predictive modeling is important, because we can't experiment with planet Earth directly! We can't make it rain for a hundred days straight over Bangladesh or cause a volcano in Iceland to erupt, just to see what might happen. We can't slowly melt the ice in Greenland and check out what happens in Europe three years later. So instead scientists create these what-if scenarios on their computers. Doing this kind of work involves a lot of different skills: computer coding and programming; an understanding of physics and chemistry to accurately capture how water and air heat up, cool down, and move around; and analytical skills to understand what the model is showing, and how that information can be used in the real world.

Habitat prediction is a type of predictive modeling where scientists study—you guessed it!—habitats, or environments. They study the ways in which specific habitats for unique plants and animals are likely to be impacted by future changes in human activity or climate. This might mean studying a particular river or mountain area that is critical to local wildlife, then modeling ways to use the land but still protect the most important parts. Habitat loss happens when humans take over land to build or pave or interrupt the natural space, and also when the changing climate causes the land to become unwelcoming to plants and animals that used to live there. (Think about an area that always had rainfall and puddles for animals to drink from, but that has been in drought or even had wildfires because of climate change. Humans might not be anywhere near that land, but the animals still have to move away.) By using habitat prediction models, scientists can prioritize the protection of lands that play important roles in sustaining diverse forms of life.

Valerie arranged funding through grants that finance these tribal-led tree planting projects. She also wrote newsletter content quarterly to communicate the work that TWP does, pursuing this nonprofit's goal of building healthy ecosystems while also creating economic-development opportunities for tribal communities. She called it "a dream job," saying that when Native kids plant native seeds, "we're not just planting trees, we're planting our future." Valerie now works with a Lakota nonprofit on the Pine Ridge Indian Reservation.

Valerie also stays active in the larger scientific community, writing papers and speaking at conferences, always seeking to contribute to the understanding of the importance of Indigenous voices in science. She thinks tribes should be able to set their own conservation goals and manage their own land, and works to help make that happen. She also helps tribes find resources to help feed people and build local economies, and she wants the scientific community to recognize the importance of using traditional ecological knowledge to guide their efforts to restore degraded landscapes within reservations.

Valerie says she loves engaging with partners, from the US Forest Service to local nonprofits to tribes, and hopes that the impact of Trees, Water & People will continue to grow and include more tribes. She says, "Our projects don't belong to us, they belong to the tribes."

Ultimately, she says, "I am trying to make sure we have the natural world full of wonderful species for my kids and grandkids and great-grandkids. I want them to have culturally and ecologically important species.

"I am trying to make sure we have the natural world full of wonderful species for my kids and grandkids and great-grandkids."

In Indigenous ways we talk about seven generations, about taking care of our land and resources thinking in terms of the next seven generations. Science and technology provide us with the tools to be proactive and commit to thinking about how to adapt to a warming climate."

As she continues to work with tribes on climate change, Valerie says she's hopeful about the future. "We try to focus on connecting tribal youth with Elders, so that Indigenous knowledge gets shared." We all need to have hope for our future, and Valerie's work helps young people in her community ask important questions about where our garbage goes, and what's in our water. It starts with asking questions and researching possible answers from Elders and through scientific research. Valerie urges all kids to ask hard questions and push for change: "You are ambassadors of the future! When you learn, you bring it home to your families and their communities. I hope readers, no matter where they live, will consider a neighborhood garden, or a school composting program, or a study to look at water resources."

IF I HAD A TIME MACHINE

What Valerie would tell herself (or you!):

"As the first in my family to go to college and get a PhD, and as a mother who went back to school, I will say it is not easy. But follow your heart, and even when there are curveballs, don't give up on your dreams. There will be people who will help you. My deep passion was to help others and lift Indigenous voices. And I learned that if you don't give up, if you keep at it, people will help you on your way. And those who don't, those who make you feel like you're not worth it, or do not respect or honor you, leave them. Value your own self-worth.

"There will be sacrifices, but for me those sacrifices were worth it. There is always another option, always a way to move forward, so don't let anyone push you into a corner and tell you you're done."

"You are ambassadors of the future! When you learn, you bring it home to your families and their communities."

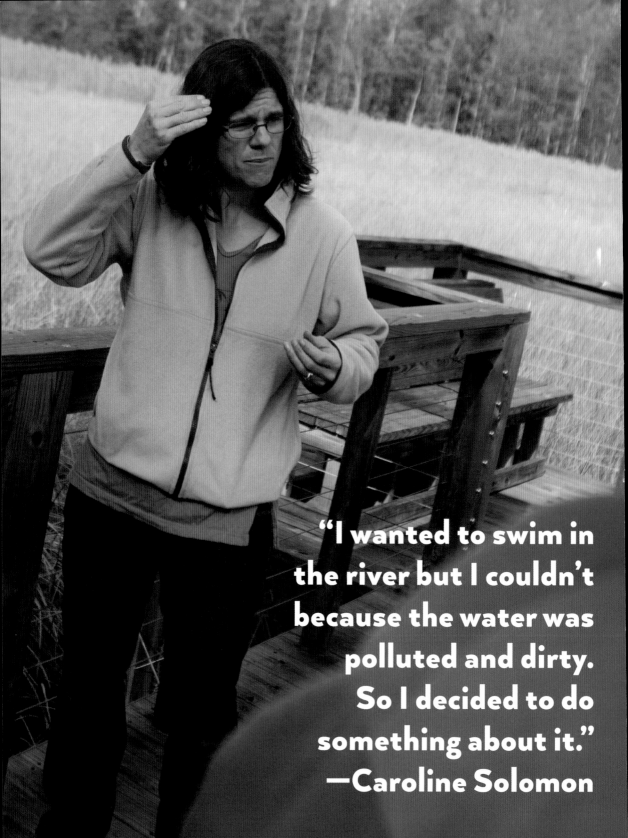

"I wanted to swim in the river but I couldn't because the water was polluted and dirty. So I decided to do something about it."
—Caroline Solomon

CAROLINE SOLOMON

For Caroline Solomon, being deaf, a gold-medalist swimmer, a scientist, a professor, and a speaker—with a sign-language interpreter—to a massive crowd at the 2017 March for Science in front of the White House are all part of her long fight for cleaner water and more inclusive science. But it hasn't always been easy. Her journey and its challenges have led to her current path, teaching ecology and biology to college students at Gallaudet University, a prestigious school for deaf and hard-of-hearing students in Washington, D.C.

Caroline was not born deaf, but survived a dangerous disease called spinal meningitis as an infant and lost her hearing soon afterward.

◀ Caroline signs to her students on a research trip near their university in Washington, D.C.

"The hardest part is not being a part of the hallway chats where new ideas and collaboration emerge."

Despite being deaf in a hearing family, a reality that can be isolating, Caroline excelled both as a student and as an athlete. She says that she always loved being outside in nature, especially in water, and liked her high school science classes, but didn't seriously consider science as a career until her third year of college. Still, she says that the seeds of her whole career were planted back in high school.

"I was a competitive swimmer, and wanted to swim whenever and wherever I could. When I was in high school we moved near Annapolis, Maryland, and there was a creek that ran right by our house. But it was too polluted—no one was allowed to swim there."

In an interview with the USA Deaf Sports Federation, Caroline said, "I became a biological oceanographer because of that. . . . I wanted to swim in the river but I couldn't because the water was polluted and dirty. So I decided to do something about it."

As a student at Harvard University, Caroline majored in environmental science and public policy, and credits a professor who offered her a summer internship between junior and senior years as a mentor who really helped her believe that there might be a place for her in science. Harvard is notoriously difficult, but for Caroline there were extra hurdles. There were no sign-language interpreters on staff when she first joined the school, though the university did eventually hire someone during her first year. But even with an interpreter, it can be hard for a deaf student to feel connected.

"The hardest part is not being a part of the hallway chats where new ideas and collaboration emerge."

MUST LOVE FISH: OCEANOGRAPHY, MARINE BIOLOGY, BIOLOGICAL OCEANOGRAPHY . . . WHAT'S THE DIFFERENCE?

Oceanography is the study of oceans, which can include their chemical makeup, the physics of their waves and currents, the ecosystems that live within them, the effects of climate change, and more. There are many different specialties, and they encompass all kinds of variations. Some oceanographers study the largest animal on Earth (the massive blue whale), and others study invisible organisms. Some focus on ocean circulation, while others study beaches. Some study populated coastlines, and some focus on deep-sea exploration. Some research the history of our planet by finding a series of clues in the ocean, and some study how marine plants and animals might have lifesaving properties when used in human medicine. There are so many different paths, but they all go through the ocean.

Biological oceanography is a branch of oceanography that focuses on the study of life in the oceans. This includes all marine animals, but also plants, microbes, and anything else that's alive in the water. These scientists often focus on how these various life-forms interact with the ocean as a whole.

Marine biology is another branch of ocean science, but one that focuses more specifically on marine organisms, and less on the larger relationship that these various plants and animals have with the ocean as a whole. They are less likely to be studying the physics of waves or the chemistry of the water. Marine biologists are often responsible for seeking out new species in the oceans, and for assessing and sharing information about populations at risk of extinction.

Caroline says, "I had a period in graduate school when I was at the University of Washington where I felt discouraged, and like I was not as smart as the other students. I tinkered with the thought of becoming a newspaper editor, or really, anything different from science."

During this hard patch she asked for time off from her graduate studies and took an opportunity as a teaching assistant at Gallaudet University, a renowned school for deaf and hard-of-hearing students. She says, "Once I landed on the Gallaudet campus, I didn't look back. I met many other deaf people who were also working on their doctorates. I finished my master's degree at the University of Washington, then moved across the

"Once you believe you can do anything, the opportunities are just everywhere."

country to take a professor job at Gallaudet. I then continued my doctoral studies at the University of Maryland."

As a professor, Caroline works directly with deaf and hard-of-hearing students, doing everything in her power to lead by example and show them that anything is possible. She is eager to train new deaf scientists and teachers, encouraging them to aim high and pursue their dreams. In an interview with National Public Radio, Caroline said, "One thing about deaf people is, very often they're isolated. They're not part of the family conversations at the dinner table, or they're mainstreamed in a public school all by themselves as the only deaf person. . . . You just have to work to change that perspective they have about themselves. Because once you believe you can do anything, the opportunities are just everywhere."

Much of the science that Caroline and her students do relates back to that polluted creek in Annapolis, not far from the Gallaudet campus. Now she studies the Anacostia River, which runs right through Washington, D.C., past

Caroline and her students prepare for an experiment at the Smithsonian Environmental Research Center in Edgewater, Maryland. In the first photo she is signing "know"; in the second photo, "take advantage of"; and in the third photo, "It looks like...".

trees and forests as well as busy neighborhoods, golf courses, traffic-clogged interstate highways, and more. Though it's been years since Caroline was a teenager wanting to swim in her backyard, many bodies of water are unfortunately still very polluted.

Caroline and her students go out in boats to collect water samples and use probes that test for oxygen, salinity, and chlorophyll. They bring the samples back to their lab and analyze them for single-celled plants like algae, and microscopically small crustaceans, which she describes as "tiny shrimp."

She says, "We can see algae under the microscope and figure out how much nutrients such as nitrogen and phosphorus are available to support it. All this helps me figure out if the water is healthy or not."

Right now many rivers are not in good shape. One problem occurs when too much fertilizer from nearby lawns and gardens runs into the river during rainstorms. These nitrogen-rich fertilizers feed the algae in the water and

Caroline explains to students how to use a YSI probe, which can collect data such as salinity, temperature, and dissolved oxygen in water.

"For much of my career I am in the lab, doing fieldwork, in the classroom, or in my office. I love how I am never in the same place."

cause its growth to explode, upsetting the delicate balance of the river. This can affect the health of the whole area, impacting everything from insect to frog to fish to bird populations, all the way up the food chain. Caroline says, "I work to improve the water quality of the waterways [the public] interacts with every day."

Climate change has impacted Caroline's work. For areas such as Washington, D.C, that are experiencing more rainfall, it means more runoff into the river—bringing more stormwater and sewage, which in turn brings more nutrients. She says, "Studying the river has a great influence on understanding how the ecosystems are changing." And while the science of climate change is happening, the commitment to understanding and fighting it depends on the politics of the moment. "The government has a large influence on what I do," Caroline says, "because they are my largest source of funding for my work. If the administration supports and believes in science, my work flourishes."

Caroline's work on the Anacostia River and the larger Chesapeake Bay area has been published in scholarly journals and won her acclaim as a scientist. But she considers the other part of her job—being a teacher and role model to deaf and hard-of-hearing students—equally important.

"For much of my career I am in the lab, doing fieldwork, in the classroom, or in my office. I love how I am never in the same place."

WHO PAYS FOR SCIENCE?

It's obvious that humanity benefits from major scientific breakthroughs, like lifesaving medicines or more efficient energy sources. But it takes years of research and millions of dollars to conduct scientific experiments. And of course there is no guarantee that scientists will get the outcome they want. So where do they get the money? Who pays for all this science?

The answer is that we do. While some projects are funded by nonprofit organizations or even corporations, most funding for scientific experiments comes through government grants. And the government gets its money, in large part, through taxes, where working people pay a small amount of the money they earn to the government to help pay for everything from schools to highways to internet service to scientific research. So as a society we invest tax money to pay for science, but it's more complicated than that. Because the next question is: Who decides how much to spend on science, and what's most important to study?

Sometimes reports of wacky-sounding science experiments make the news, and some politicians use these stories to complain. Government money—our tax dollars—spent researching shrimp on a treadmill, or the amount of nicotine found in toenail clippings? Sounds like a waste of money, right? Except that the shrimp experiment was part of a study on the sustainability of shrimp in different conditions, which matters a lot to the shrimp fishermen who count on them to earn a living, and to the communities where they live. And the toenail-clipping experiment was part of a project to help understand people's risk of lung cancer. Often these news headlines are intended to make voters angry about paying their taxes, but they don't tell the whole story.

Depending on who is elected president and to Congress, the priorities around science funding can change dramatically. This makes it hard for scientists to plan their research. Still, regular people do have an opportunity to make their opinions known! You can write your elected officials and tell them you want them to support scientific research. You can raise money (Lemonade stand! Yard sale!) and donate it to a nonprofit organization that funds science.

Also, not all grants are from the federal government, and not all of them ask for millions of dollars. Did you know that your teacher can write a grant request for a small amount of money for your class to do a science experiment? Or that your neighborhood organization, mosque, church, or synagogue can request a grant for a local garden or park? Cities and towns, nonprofit organizations, and other groups give money to help kids and schools do more science and improve their environment. And who knows what new discoveries might get made . . . right in your backyard.

Caroline loves working with deaf and hard-of-hearing students in part because she believes that these young people need more representation. In the interview with NPR she said, "Growing up [as the only deaf person in many of my classes], what I needed was role models." She continued, "Our [deaf] students are so very visual. You have to think about everything in a visual way. . . . I wish someone would have taught me like this when I was in college. Often, faculty are so focused on teaching from their own perspective, they're not thinking about the student side of things."

Beyond the walls of her university, Caroline has worked to expand opportunities for all deaf and hard-of-hearing students in the sciences. She helped create the ASL-STEM (American Sign Language–Science, Technology, Engineering, and Math) Forum, which is building a standardized set of signs for use in scientific and STEM settings. She also led a workshop for deaf and hard-of-hearing students that was supported by the National Science Foundation and explored ways to better encourage and support these students in the sciences. In that workshop she highlighted important contributions that deaf and hard-of-hearing scientists have made throughout history, and pointed out that increasing the diversity of people practicing science leads to better outcomes.

> "I would tell them to be curious! To experiment and play and not be afraid to get dirty."

For Caroline, urging deaf and hard-of-hearing students to consider careers in the sciences is an important part of her job.

"I would tell them to be curious! To experiment and play and not be afraid to get dirty." She also encourages all kids to recognize that the subjects they're learning now will be useful later in their careers. "I use algebra, trigonometry, and calculus all the time!"

But most of all, she wants kids to recognize that we all need to be working to help our planet. "There is a lot we can do!" she says. "The first is simple: reduce the amount of trash we produce, and make sure we dispose of it properly. I see a lot of trash in the river. But also kids can advocate for their families to put solar panels on their houses, or use rain barrels to collect rainwater to water their gardens. I want people to know that it's not too late to reverse the trends that are destroying our planet, but we have to move quickly!"

AMERICAN SIGN LANGUAGE AND SCIENCE ARE GROWING TOGETHER

We don't often think about where words come from. We might have to think about different languages, though. Some of us know more than one language, and some of us learned one language at home with our families and another at school. Some of us have parents who can't easily understand our teachers, or have grandparents who speak a language we don't understand. Translating things from one language to another can be tricky, and sometimes the meaning gets lost in translation.

For deaf and hard-of-hearing folks in North America, **American Sign Language**, or ASL, is the most common language they use to communicate. Instead of speaking, they use their hands and facial expressions to signal the words. (It should be noted that there are hundreds of other sign languages all around the world!) While ASL has thousands of signs, there are many more words in English than there are ASL signs to match. And in science, technology, engineering, and math, often called STEM subjects, new words are often created as new techniques are developed or scientific breakthroughs occur. In STEM, having the precise word to explain what is going on is really important, and language changes to adapt to new concepts. But while scientists might agree on a term in English, there is no translation in ASL.

That's why the **University of Washington's ASL-STEM project** is so cool. Its goal is to connect thousands of ASL users working in STEM by adding new scientific terms that can become part of the larger vocabulary. By adding to the ASL-STEM project, Caroline and her students are helping grow the language for the work they're doing in science!

"We want to make sure plants get their fair share." —Rupert Koopman

RUPERT KOOPMAN

Rupert Koopman, a botanist by training who works with the Botanical Society of South Africa, grew up surrounded by what he calls "plant-y people"—people who know and love plants. His father's family ran a nursery, and his dad grew up learning about and working with those plants before becoming a science teacher. His mother's family lived in a rural area where, Rupert says, "They grew squash and maize and flowers and all sorts of things, and whenever we went to visit I would get lost in the gardens forever."

This early exposure to some of the native plants of South Africa made a difference. Rupert explains, "My dad knew that **fynbos**, the local vegetation,

◀ Rupert's love of the flora of South Africa makes him a passionate and outspoken advocate: he is a voice for the plants.

WHAT IS FYNBOS, AND WHERE DOES IT GROW?

The province where Rupert lives in South Africa is the Western Cape, which is renowned for a group of plants collectively called **fynbos** that cover much of the landscape. The region, known as the Cape Floristic Region, is one of the world's six floral kingdoms and is the only one that is contained within a single country. While some fynbos plants might look uninteresting or dormant much of the time, they bloom with an amazing array of different flowers and survive in unforgiving soil that often won't support many other species. The fynbos biome is home to more than nine thousand plant species, with almost 70 percent of those being unique to the land. So while the vegetation looks ordinary to the untrained eye, to someone like Rupert, who grew up learning about local plants, fynbos is something very special.

These are just a few of the vast array of plants that grow in the fynbos biome.

"It hit me right between the eyes: this is a community I wanted to be a part of."

was something really special. It doesn't grow anywhere else in the world. So from an early age I had someone to help me interpret what I was seeing. And it's important: it allows you to zoom in beyond that area of amorphous green that, to the untrained eye, all looks the same.

"Unless you have access to plant-y people, I don't think you grow up learning to look carefully."

When Rupert went to university he decided to study science, but, as he puts it, "I didn't know where it would take me. I took a long time—six years to complete what could be a three-year program—because I wasn't sure what I was looking for." Ultimately Rupert finished his degree in botany and geography, which he says "was a really nice balance, because they are two complementary areas, looking at the ways that landscapes affect plants and vice versa."

It was an ecology lecturer who heavily influenced the path of Rupert's career. In his class, Rupert had the chance to do fieldwork and put together a project in a nearby nature reserve, which is part of the fynbos region. Rupert was selected to attend the Fynbos Forum, a regional conference for all kinds of people who work in the area, including conservation experts, professors, researchers, and more—from the people who wrote the textbooks he was studying to people who spent every day in the field. Rupert found himself surrounded by folks who all cared about this unique region.

"It hit me right between the eyes: this is a community I wanted to be a part of."

He continues, "That conference got me over the hump of not being sure what I wanted to do with my life, to really knowing that I wanted to be involved in this. Most of my career has been born from that conference."

Rupert began working with a group called CREW, Custodians of Rare and Endangered Wildflowers, which is still running today. He says, "We have so many threatened and unique plants in South Africa, and not nearly enough state-employed botanists to do the work of monitoring them. So CREW recruits community or citizen scientists, trains them, and sends them out to do the monitoring and gathering of data."

Land use and conservation in South Africa is tangled up with the country's history of apartheid. Apartheid was a system of legal racism that forced Black South Africans, as well as those of mixed race, to live separately and very unequally. White South Africans ruled the country and made all the decisions about what land was conserved and who was allowed to use it. The law changed in 1994, but even years later there are still many parts of the country where inequality persists. Land conservation and land use was just one area where Black South Africans were not well represented. But that is starting to change. When Rupert began his career, organizations had started offering short-term job contracts to attract diverse candidates who might not have thought about a career in conservation or plant management. Colleagues from CREW and CapeNature wrote an application for a botanist, Rupert applied successfully, and that became a stepping-stone into CapeNature, the organization where he spent more than twelve years.

Rupert realizes that he was lucky to find a path into this work, and part of his career has been building connections to help others follow in his footsteps. He remained connected to the Fynbos Forum, and spent several years as the chair of the conference where he made his first connections.

During Rupert's time at CapeNature he served as a botanist and resource ecologist, which he describes as "a dream job." He had a pivotal role in monitoring some of the most threatened and unique plants in the world, and his work ranged from assessing threats facing vulnerable species to acting as a "voice for the plants" in development and construction conversations,

> "I provided an extra layer of botanical eyes."

interacting with landowners, developers, and public land trusts to ensure that the fynbos had an advocate. Much of his time was spent in the field, traveling around Western Cape province to monitor the health of plants, assess new dangers, and show up for public hearings.

He says that his job was a good mix of fieldwork and science communication. One of the interesting parts of the work was finding hidden species in the landscape, and documenting where they were growing and where they were disappearing. But another part, which he describes as equally important, was letting people know what was happening, and providing an "extra layer of botanical eyes." He says, "One important thing to remember about many fynbos plants is that they're seasonal. So land assessors might send someone out to look at a piece of land, and it's the wrong time of year, so they miss a lot of interesting plants. Then they produce maps that say there's nothing out there. And I would take a closer look and say, 'Hey, you missed the important stuff! Just wait until later in the year!'"

Rupert reiterates that the secrets of the fynbos region are unique to that place. Once, he said, a well-respected botanist had observed a piece of land for farmland assessment, and said there was nothing important growing there. But Rupert, who had studied that general area for more than ten years, knew this was wrong. He and colleagues suggested that the project managers wait until the first rains arrived, and sure enough, they found an *Oxalis*, a genus of plants with many unique local species, that hadn't been seen in that area for over sixty years. Often his job involves trying to convince developers to slow down. He says, "The problem is that developers are naturally in a hurry, trying to push projects through, and there is a need to observe the land in peak flowering time before making final decisions."

That's why Rupert is so excited about the army of volunteers, of citizen scientists, who are helping map the land. Rupert and other experts provide training, but there are nowhere near enough experts to do all the fieldwork that's required, so volunteers fill in the gaps. New technology like the iNaturalist app, which acts as a virtual repository for all the data that volunteers collect, is making a real difference. "We get a pipeline from the volunteer in the field directly toward conservation efforts and protection of the plants."

The work Rupert does with plants relies on both old-fashioned tools and cutting-edge technologies. He says that one of his main tools is the plant press, which goes back to the Renaissance or even further. A device that is simply two pieces of wood, straps, and pieces of paper, it flattens and "presses" a plant so that it can be preserved. At the other end of the spectrum, satellite technology and Google Earth have dramatically changed the way he helps build prioritization maps of the land. It is much easier to get a detailed look at the land and what's growing there via satellite before trying to travel to every corner of the region. "We can now zoom in to an area the size of my house, where it used to be that we were looking at a fifty-square-kilometer patch and hoping for the best."

In 2020 Rupert left his regional role to join a national nonprofit organization, the Botanical Society of South Africa. Where he used to focus all of his attention on one province, Rupert is now responsible for guiding the conservation strategy for an organization that serves the whole country. He spends less time in the field practicing botany, and more time working toward partnerships with other organizations, trying to educate the public and Botanical Society members about why conservation matters. "We want to make sure plants get their fair share."

Much of his job is storytelling about the importance of conservation in

"I like to look across the landscape of South Africa and ask how we can help plants get a better deal than they're currently getting."

general, and the unique landscape of South Africa more specifically. While the National Botanical Gardens are rightly famous, Rupert wishes more people knew how important it is to protect plants in the wild. He says, "I like to look across the landscape of South Africa and ask how we can help plants get a better deal than they're currently getting."

The great challenge is that in South Africa, as in most of the world, conservation is too often seen as a luxury. Rupert says that "far too often if there was a choice between new farmland or a new development, and conserving the land, the support goes to the option that leads to economic benefits." But in Rupert's mind, that is short-sighted. "These native plants, many of which are found in and just outside of major cities, are some of the most unique in the world. And there's no way to explore how special these plants are before they disappear. We know that plants in remote jungles and rain forests can have important medical benefits, and the same is true here. Just recently a researcher in Stellenbosch, just outside of Cape Town, walked in the mountains in the fynbos region, scooped up some soil, and analyzed it. He found that there were natural antibiotics in the soil that showed potential against multi-drug-resistant strains of staphylococcus bacteria."

Rupert refers to conservation as "the art of keeping your options open," recognizing that the more natural plants and land that remain, the more options the planet has to adapt to climate change. But the effects of

"Climate change shifts the urgency of the situation, making it more important than ever that we protect the most fragile places."

Growing up exploring the outdoors was a big part of Rupert's upbringing, and one he shares with his daughter.

climate change are accelerating the loss of biodiversity. In South Africa, a long-term drought a few years ago put the city of Cape Town within thirty days of running out of water. In desperation, the city reversed many environmental protections, allowing people to drill new wells in sensitive areas and wetlands. But the climate crisis, Rupert believes, should not be an excuse to destroy more of the landscape—just the opposite.

"Climate change shifts the urgency of the situation, making it more important than ever that we protect the most fragile places."

When talking about how kids can get involved, Rupert is clear that there are plenty of options. "There are lots of forms of activism, from supporting local natural areas to holding the adults around you accountable." He continues, "Far too often people who are messing up the environment for economic gain say they are doing it to make a better life for their children. But the children need to speak up! And make it clear that what they really need is for people to protect the planet."

RUPERT'S SURPRISING ADVICE TO STUDENTS WHO LOVE SCIENCE
"You have to have a good grasp of language to communicate the importance of this work."

Ecology and botany can involve math, chemistry, computer coding, and more, but Rupert says that of all his high school classes, the one that surprised him with its usefulness was English class. He urges kids interested in the sciences to also work on their communication skills. There is much more writing and storytelling in his job than he expected, and it matters a lot. If you can't make people understand why the work you're doing matters, if you can't convince people to care about the land and the planet, then all the science in the world won't make a difference.

He says, "One of the great risks we have here in South Africa is that we haven't gathered sufficient knowledge, so that when older botanists die, we lose their wisdom. Part of my job is working towards systems for capturing and communicating their stories."

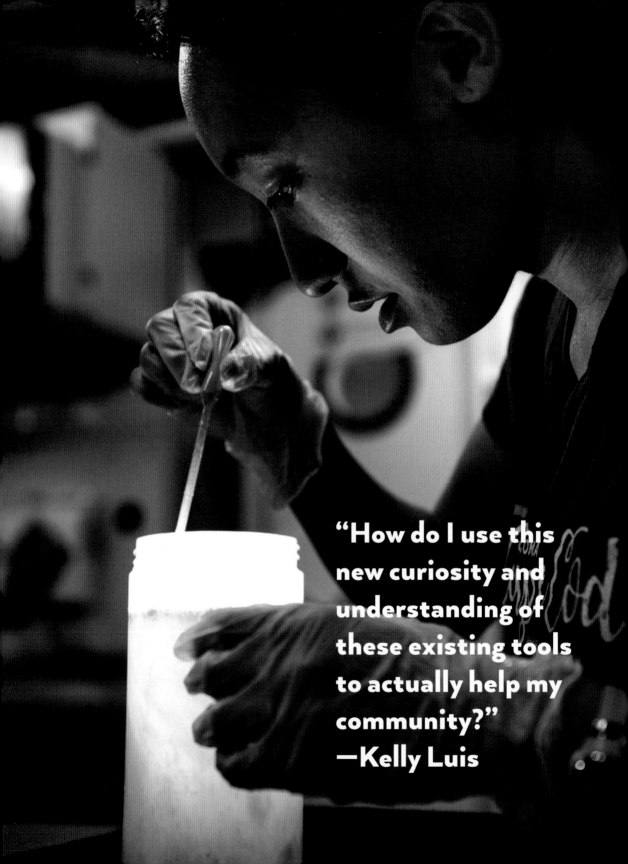

"How do I use this new curiosity and understanding of these existing tools to actually help my community?"
—Kelly Luis

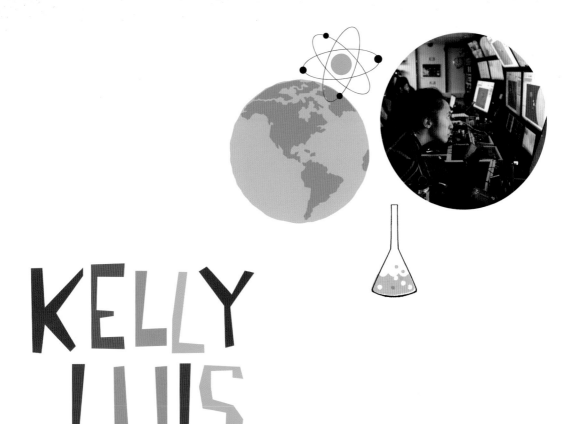

KELLY LUIS

Kelly Luis finished her PhD in marine science and technology at the University of Massachusetts Boston, a long way from her home on Maui, Hawaii. While the urban center of Boston is very different from the beaches of Maui where she grew up, Kelly sees the connections. She developed a class for university freshmen on place-based studies. She says, "UMass Boston is unusual, because it's an urban campus but it's surrounded by ocean. I thought for these students who are arriving at the UMass campus from all over, it would be cool to help them understand their surroundings—not just the science of the place, but the communities that inhabited the areas as well."

◄ Onboard the research vessel *Falkor*, Kelly takes samples of Trichodesmium, a type of algae, with a pipette. The container also housed many tiny jellyfish, no larger than a quarter.

This intersection between ocean science and humanity is at the root of Kelly's interests. Though at first glance her area of study seems strictly academic, Kelly's interest in oceans is intertwined with her interest in people. She says, "I never met a scientist, growing up. I might have had an image in my mind of someone in a white lab coat, but I never met one." She laughs. "Even my mom was surprised when I went into science, because she says that every photo of me as a kid on the beach I'm just sitting around, I'm not poking at anything or off exploring."

Kelly says that while she went to the beach almost every week when she was a kid, she was never particularly interested. "My brothers dove and caught fish, but I would just hang out. I had never heard of an oceanographer, and it certainly wasn't on my radar to become one."

But while science wasn't something she was encouraged to pursue, Kelly says that she was always made aware that education and school was her path to more and better choices in life. Her blended family has seven siblings, and she knew her role was to take advantage of educational opportunities to contribute to her community. Kelly was especially interested in politics and economics, and saw these subjects as tools she could use to give back. Kelly did well in high school and participated in College Horizons, a college readiness program for Native Americans, Alaska Natives, and Native Hawaiians, which helps students prepare for college, from assistance in getting application fees waived to helping them find financial aid. Through the program she met with a recruiter from Columbia University in New York City. Ultimately, Kelly said, prestigious Columbia University, one of the most competitive schools in the country, met 100 percent of her family's financial needs, so it would cost her less to go there than to her local state school.

Kelly acknowledges that it was hard to move so far away, to a place she hadn't even visited before, but, she says, "I was always hungry to explore beyond my island. I knew I needed to move away from Hawaii and figure out what I could bring back if I wanted to improve the situation from which I grew up."

Kelly worked three jobs while she was a student, and though it was challenging, she says, "Growing up I truly understood that pursuing my education was not a luxury; it was a means for survival. We are all dealt certain cards in life, and I had the privilege to go to this university, and I knew I owed it to myself and my family to try my very best."

The summer after her first year of college, Kelly worked as an **intern** in the Washington, D.C., office of Mazie Hirono, current U.S. Senator for Hawaii. And that's where her path intersected with science. "I was doing typical intern things like answering calls and listening to constituent concerns, and what I quickly realized is that everyone was calling about environmental issues! They were calling about something happening on their land, or something happening on the shore, or in the ocean. And that was my realization: if my community really cares about this, I should take an environmental-science class."

Kelly returned to Columbia and took an earth-science course, thinking that it would give her some useful information for her political career. But, she says, there was a

WHAT IS AN INTERN?

An **intern** is someone who works at an **internship**, which is a short-term work experience offered to students and other people seeking to learn about a career. Very often interns are college students hoping to gain experience and learn about a career path. Internships can be paid (Yay!) or unpaid (Boo!), and usually involve low-skill tasks that require no experience but that offer a chance for the intern to experience firsthand what it's like to work in that particular field. The goal of the internship is to get some hands-on experience and to see if it's a good fit. Working as an intern can help you get a job later in your career, as it offers a chance to put something on your résumé and network with people. But being an intern is also a great way to decide if you even like the career! Sometimes after an internship you might decide not to pursue a certain career path, and that's useful too.

moment that changed everything. "We were learning about types of lava in class. And there are two terms scientists use that are actually Hawaiian words: pāhoehoe, which means smooth lava, and ʻaʻā, which is a harder, crumbly lava. And there was something about a part of my language being embedded in science—it was a small thing, but it opened the door. That's when I thought there might be a place for me in science."

As Kelly progressed toward her science degree she dug into a variety of different areas in environmental science, from studying rivers to assess their chemistry to analyzing deep-sea sediment cores. All of these experiments examined the chemistry and geology of past oceans, and used those markers to understand what happened to the planet long ago. "I realized there are all these scientific tools we can use to find out what the levers were that moved our past Earth, and that they could then tell us about what might be happening with climate change." For Kelly, learning about this body of knowledge was "like a wave flooding over me!"

She says, "There were two main feelings. The first was the realization that there is so much I don't know! But the other was the sense of 'Wow, there is so much we can use.'" As a Native Hawaiian, Kelly says, she is from

"The more I learned, the more it led me to ask: How do I use this new curiosity and understanding of these existing tools to actually help my community?"

a community that is underserved by science and in many other ways.

As she pursued her studies, Kelly knew she wanted to get into a branch of science that let her use the available tools not just for academic pursuits, but also to help marginalized and Indigenous communities. "The more I learned, the more it led me to ask:

PICTURES FROM THE SKY
One of Kelly's favorite places to see satellite images of what's happening is NASA's Earth Observatory website. Its "Image of the Day" shows a cool new satellite image, with a link to learn more about it!

How do I use this new curiosity and understanding of these existing tools to actually help my community?" She quickly became fascinated with satellites, specifically those that take images of oceans and coastlines from space. She says, "I realized that these satellites that are in the sky, that are paid for by taxpayers like me, all share their data freely. So even as a kid in Hawaii, if I had wanted to, I could have looked at these beautiful images taken over the ocean, and started asking questions: What makes that pattern? What causes the water to change color? That access was there and I didn't even know it."

Ultimately, Kelly describes what she does now as studying the color of the water to understand what that color can tell us about what's happening in the ocean. She does this by using a variety of tools mounted on satellites and airplanes and boats.

When Kelly first started out in the field, she quickly realized that she would need to understand computers and learn to write computer code if she wanted to work with satellites, and the more she learned, the more she enjoyed the work. "What might surprise people," she says, "is that even though I'm an ocean scientist, I spend most of my time at my desk, writing code for hours!" The branch of science she studies, satellite oceanography, studies ocean color. She explains that this involves optical theory—how our eyes see color—and also physics and light refraction. The satellites can be built and programmed to look for different data, and to ask different

"There's a moment when I feel like a painter, when I am studying these images and decided how to use color to help tell the story."

Kelly (front) studies data in the control room onboard the research vessel *Falkor* somewhere near Fiji. She's studying the results coming in from sensors attached to unmanned aerial vehicles that were launched off the boat.

WHAT ARE PHYTOPLANKTON?

Phytoplankton are microscopically small organisms, mostly green algae, that grow in fresh and salt water. They play a vital role in the food web, so the fate of many species—and even your favorite swimming spot—depends in part on the health of the phytoplankton. Human activity and climate change can greatly impact phytoplankton, as warming seas, pollution, and agricultural runoff change the delicate balance of micronutrients in the sea. While scientists take water samples and study them under microscopes to learn some things about phytoplankton, satellite images like the ones Kelly studies offer another perspective on what is happening in our water.

questions, from finding **phytoplankton** or algae blooms in the ocean to understanding what's changing in fishing grounds. As Kelly describes her work, it's clear that it is both science and art: the biology and physics and computer coding involved in finding out and interpreting what is happening in the water, but also the creativity of deciding what questions to ask, and the coloring of the images to best highlight the information. Kelly says, "There's a moment when I feel like a painter, when I am studying these images and decided how to use color to help tell the story."

Kelly says that she spends a lot of time "fighting or dancing with my code—sometimes I can't get the code to run, and I will get an image that I know isn't right. It's showing something but I know it isn't true, based on measurements people have already taken. And I start poking and wrestling, and sometimes I have to put it away and rest my mind."

But not all of Kelly's work is at a desk. She is clear that while her beloved satellites can do amazing things, there is still a vital role for humans as well. In fact, one of the most important jobs of the satellites is to help the human scientists know where to go. The ocean is enormous, and without a hint where to send a boat to gather samples, many of Kelly's projects would be impossible. While in graduate school, she says, they would regularly go out on what were called "satellite validation" trips for NOAA, the National Oceanic and Atmospheric Administration. She and her fellow scientists would go out into Massachusetts Bay and confirm what the satellites were reporting. "Those were fun days," she says, "because it a ways depended on the weather, so you might wake up and have to rush off to get ready and get on the boat and head out to sea. And we could be out there for hours, assessing what we're seeing in the water and what the satellite picks up. Then we would get back to shore, and bring our water samples to the lab and analyze them, and often it was late at night before we finished." Those days, mixed with her days focused on computer code, provided a good balance.

Much of Kelly's work is solitary, sitting at her computer and either writing code or viewing images. But at the same time, Kelly says she feels like part of something larger. "There's a sense of being a single puzzle piece, but there's also a real sense of camaraderie with other satellite data users, because there is a whole team working on applying satellite data to address big environmental problems in our own pockets around the world."

While many of the tools Kelly uses are cutting-edge, she also uses techniques that go back hundreds of years. One of these is the Secchi disk, a tool created by astronomer Angelo Secchi in 1865. Essentially it is a standardized black-and-white disk that is lowered into the water on a string, and the depth at which it disappears from view illustrates how much light is able to penetrate the water column, which in turn can show how clear or murky the water is. The tool being used now is very similar to the original, but Kelly also talks about how one of her colleagues is using satellite technology to get standard Secchi disk estimations of coastal inland water bodies from space.

Whether she's on a boat or at her desk, Kelly is clear: the choice about what to study, and what satellites to build, should depend on what people and communities want to learn about, and this should be based on what vulnerable populations need. She says that often what drives their work is that people living on the coasts, the people actually living with the water she's studying via satellite, have questions. "Sometimes they are noticing more algae blooms in the water, and it's just common sense: you don't want to swim when the water is murky and yucky. Or sometimes there are less fish than there used to be."

Most of these trends, she says, are connected to our changing climate. One example she gives is that of a major fishing area in New England. There are little eels that are very common there. People don't fish for them, or care very much about them, but while doing fieldwork in New England, Kelly learned from fisheries scientists that the eel population was moving farther north up the coast as the water warmed, because they preferred cooler waters. And many species of fish eat these eels, so they followed the eels north. Larger fish, and even whales that eat those fish also moved north, leaving parts of the fishing ground emptier. And while Kelly says that satellites can't see all the way down to the bottom of the sea, they can show us what's happening at the surface, which can offer hints of what's happening down below and provide a map for scientists of where they should look for more information.

The subject of climate change can be a challenging one, and Kelly learned that different political administrations can mean very different levels of funding and prioritization for environmental work. But she says that wanting water to be clean—for swimming, fishing, or living near—is not a political issue. "When I talk to people, no matter where they are on the political spectrum, I try to remember that everyone wants the quality of their water to be good."

She also urges everyone—kids and adults alike—to get involved. There are several citizen-scientist projects that let people monitor water

> **"What I really want everyone to know is that no matter where you live, no matter who you are, you are in a relationship with what is happening in the ocean. The air we all breathe is linked to the phytoplankton and everything else that lives there."**

quality in their communities. Kelly says, "Anyone with a phone can download one of these apps and record water quality at their local lakes and [other coastal] areas. The EyeOnWater app allows users to take a photo of their water body, and the data is contributed to a global water-quality database. And that information becomes really important to the work I do."

"What I really want everyone to know is that no matter where you live, no matter who you are, you are in a relationship with what is happening in the ocean. The air we all breathe is linked to the phytoplankton and everything else that lives there. And maybe you weren't part of the historical decision making that led us to this point, but you are here now."

Kelly also feels strongly that everyone is in a position to be a part of the solution, and to be a scientist, whether officially or not. She says, "I've learned so much from my Hawaiian family, in terms of the value of good observational skills and the ability to ask the questions that help you formulate ideas on what's happening around you. These are skills scientists use, but all school does is further sharpen them. We all have these skills already."

"If I can demystify one thing for kids, it would be that all being a scientist means is that you are a person who truly focuses on building those observational skills in different settings."

For example, Kelly says, "My brother knows more about the impact of sunscreen on coral reefs in Hawaii than most scientists, and can have conversations with scientists at high levels, even though he hasn't taken a lot of science classes." She continues, "If I can demystify one thing for kids, it would be that all being a scientist means is that you are a person who truly focuses on building those observational skills in different settings."

Kelly also wants to remind future scientists that "it's okay not to know things! It's okay to ask questions! Part of being a scientist is getting comfortable with not knowing things, and realizing that you have a right to ask questions and have someone explain things to you." She laughs. "It took me a long time to realize that just because I have to ask a question, just because I don't know, that doesn't mean I'm stupid. Your questions are worthy."

SUNSCREEN CHEMICALS AND MARINE LIFE

How sunscreen chemicals enter our environment:

The sunscreen you apply may not stay on your skin.

When we swim or shower, sunscreen may wash off and enter our waterways.

How sunscreen chemicals can affect marine life:

Chemicals in some sunscreens that can harm marine life:

- 3-Benzylidene camphor
- 4-Methylbenzylidene camphor
- Octocrylene
- Benzophenone-1
- Benzophenone-8
- OD-PABA
- nano-Titanium dioxide
- nano-Zinc oxide
- Octinoxate
- Oxybenzone

GREEN ALGAE: Can impair growth and photosynthesis.

CORAL: Accumulates in tissues. Can induce bleaching, damage DNA, deform young and even kill.

MUSSELS: Can induce defects in young.

SEA URCHINS: Can damage immune and reproductive systems, and deform young.

FISH: Can decrease fertility and reproduction, and cause female characteristics in male fish.

DOLPHINS: Can accumulate in tissues and be transferred to young.

Here are a few ways to protect ourselves and marine life:

Consider sunscreen without chemicals that can harm marine life, seek shade between 10 am & 2 pm, and use Ultraviolet Protection Factor (UPF) sunwear.

| Seek shade | Umbrella | Sun hat | Sunscreen | UV Sun glasses | Sun shirt | Leggings |

Revised Sep. 2020

oceanservice.noaa.gov/sunscreen

This infographic from the National Oceanic and Atmospheric Administration depicts how chemicals in sunscreen can have a negative impact on marine ecosystems, including coral reefs. Scientists and surfers alike are playing a role in better understanding and protecting them.

"It's okay not to know things! It's okay to ask questions! Part of being a scientist is getting comfortable with not knowing things, and realizing that you have a right to ask questions."

There is no time too early or too late to get involved. There are no skills too small or too unimportant to matter in the fight against climate change.

Non-Scientists Who Are Making a Difference

Pro tip: you don't have to be a scientist to save the planet. And here's another one: you don't have to be an adult with a long educational career.

The scientists interviewed for this book are all—in their own ways—working on making our planet healthier. Whether they're measuring the corals in the ocean or mammals in urban settings, they are using the tools of science and discovery to answer questions about the different species of plants and animals who share their home with us.

But just as there are many other scientists studying climate change and our planet, there are also millions of other activists, artists, photographers, and politicians who are working to combat climate change and help heal our Earth. And some are younger than you, and some are older than everyone in this book.

There is no time too early or too late to get involved.

There are no skills too small or too unimportant to matter in the fight against climate change.

So whether you love hiking up mountains or surfing or writing computer code or taking photographs or speaking out in a crowd or painting murals or kicking a ball, there is a role for you.

JAMIE MARGOLIN, *STUDENT AND YOUTH ACTIVIST* (USA)

Jamie was fourteen when she first started organizing protests to raise awareness and spur action about climate change. A Colombian American living in Seattle, Washington, Jamie quickly connected with other teens around the country and co-founded Zero Hour, a youth activist organization that centers diverse young people and works to empower them to push for action. As someone who identifies as Jewish, Latina, and lesbian, Jamie feels strongly that social justice, climate justice, and care for our planet must all be woven together. She believes that young people can use their power in numerous ways—protesting, writing to their elected officials, starting community projects, and more. Her first book, *Youth to Power: Your Voice and How to Use It*, came out in 2020.

KATHY JETÑIL-KIJINER, *POET AND ACTIVIST* (MARSHALL ISLANDS)

Kathy is from the Marshall Islands, a small, independent island nation in Micronesia. While Indigenous people have lived there for thousands of years, the land is increasingly vulnerable to sea-level rise due to climate change, and was damaged by US nuclear testing that took place there in the 1940s and 1950s. Kathy uses her poetry and art to educate people around the world about the dangers of climate change, the importance of Indigenous voices, and the interconnectedness of our planet and our people. In addition to addressing a global audience of thousands for the United Nations climate summit with a poem written for her seven-month-old daughter, Kathy has founded a nonprofit, Jo-Jikum, which means "home" or "your place" in Marshallese. Through her poetry and her activism she works to inspire people to turn the tide on climate change.

INKA CRESSWELL, *UNDERWATER FILMMAKER AND DIGITAL STORYTELLER* (UK)

Inka would always rather be underwater, a fact that's been true since she learned to snorkel as child, and only grew as she became a certified scuba diver at eleven. She was born in the United Kingdom, traveled to California to study marine biology, then decided that the best way to save the oceans she loves is by helping tell the stories of the extraordinary environments that are at risk. She has filmed all kinds of sharks, from white sharks to hammerheads to whale sharks, dove with dolphins in the Bahamas, and worked to capture the beauty of life on coral reefs around the globe. Inka still uses the knowledge gained through her science degree, bridging the gap between scientific jargon and the general public and helping communicate the urgency of ocean conservation. She hopes that by showing up in the largely male-dominated space of wildlife photography and ocean science she will inspire other girls to find their way into these fields as well.

LESEIN MUTUNKEI, *STUDENT, FOOTBALLER, AND TREE PLANTER* (KENYA)

Lesein was only twelve years old when he first learned about the deforestation happening in his home country of Kenya. Deforestation around the globe is the second-largest contributor to climate change, and has negatively affected the outdoor environment around Lesein's home in Nairobi. He decided he wanted to do something, so he channeled his love of soccer (called football in Kenya and most everywhere except the United States!) and started planting a tree for every goal he scored for his team. This initiative grew into Trees for Goals, which now has been adopted by schools and clubs around Kenya. The plan is simple: plant eleven trees—one for every player on a team—for every goal scored in practice or games. Kenya's forestry manager helps groups find forest areas to plant in, and gives advice on what kinds of trees will thrive. So far over fifteen hundred trees have been planted, and Lesein, still a teenager, has been recognized globally for his work. He hopes that soccer teams around Africa and beyond will connect with Trees for Goals . . . and that maybe even professional soccer and the worldwide organization FIFA will get involved!

RHIANA GUNN-WRIGHT, *CLIMATE POLICY ADVISER* (USA)

In the United States, political debates often mention the Green New Deal, an ambitious political framework to invest in the American economy, shift it away from fossil fuels like coal and oil, and—just as importantly—combine environmental goals with social and racial justice targets that would help

Black, immigrant, and low-income communities. The Green New Deal is talked about a lot, but less well known is that one of the people who developed and wrote this policy is Rhiana Gunn-Wright, a former intern for Michelle Obama and a prestigious Rhodes Scholar. Rhiana grew up on Chicago's South Side and struggled with asthma as a kid as a direct result of pollution in her low-income community. Her interest in environmental policy and justice is a direct result of her own experience. Now she is one of the architects of an exciting political policy that could redefine how America tackles climate change and the unfair burden that poorer communities take on. Rhiana sees policy as an important tool to make sure that power is shared and that no one is considered disposable, and as a key player talking to politicians throughout the government, she is ensuring that all voices, including the planet's, are heard.

RON FINLEY, *THE GANGSTA GARDENER* (USA)

Ron has been questioning rules and getting them changed from the time he was in school. But the episode that most changed his life was when he started growing vegetables in a barren plot of land near his home in South Central Los Angeles, only to be told it was against the law. Ron fought the law and won, and since then has become internationally known for his commitment to helping people grow their own food, wherever they live. The Ron Finley Project now teaches communities that growing their own food is the most powerful, delicious, and yes, gangsta thing people can do. While Ron believes that everyone should be empowered to garden and grow their own food, he is particularly eager to see change in communities known as food deserts—low-income neighborhoods where fresh produce and healthy food are hard to find. But he's clear that everyone, whether rich or poor, Black or white, will benefit from reconnecting to the Earth and recognizing the importance of air, soil, water, and plants.

HELENA GUALINGA, *STUDENT AND ACTIVIST* (ECUADOR)

Helena grew up in the Sarayaku Indigenous community in Ecuador, in a remote part of the Amazon forest. Her community is small and fragile, made up of only around fifteen hundred people. They are one of over eight hundred Latin American Indigenous groups, many of which have experienced climate change firsthand through drought, more frequent forest fires, and earlier and faster snowmelt. Like many others, the Sarayaku are struggling desperately to keep their land out of the hands of corporations that would blast, drill, and mine the Earth for profitable oil and gas. Many women in Helena's family are active in their pursuit of Indigenous land rights, and Helena has seen the dangers that can arise when activists and governments clash. Throughout her life Helena's family has moved between Ecuador and Finland, and as she went to high school in these very different countries, she found connections between the struggle to protect her own homeland and the larger struggle for climate justice. Even as a young teen, Helena realized that her education and tools (like speaking English) give her a platform to raise awareness. Helena views herself less as an activist and more as a voice for thousands of activists—Indigenous and otherwise—who have already been doing this work across Latin America without being heard. In 2020 Helena co-founded Polluters Out, a global activist group focused on pressuring governments, companies, and banks to break ties with the fossil-fuel industry.

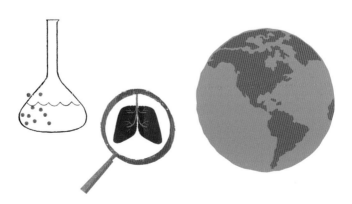

DIY ACTIVITIES

HOW TO BUILD A PLANT PRESS

A plant press is a basic but vitally important tool that botanists and ecologists use to preserve plant specimens. It involves laying the plants flat between pieces of paper that can absorb moisture as the plants dry out, with cardboard in between to allow airflow, and enough weight on top to flatten the plant. The amount of time and weight needed depends: it would take a lot more weight to flatten a giant sunflower, and a lot less to flatten an oak leaf! The basic model hasn't changed much in the past five hundred years, and you can make one at home or at school. If done correctly, the plants you preserve can last for hundreds of years!

Here's what you need:

MATERIALS
- heavy cardboard
- newspaper
- flat, heavy items like big hardcover books to use as weights*
- some kind of strap (You can use a belt or a piece of rope.)*
- heavy-duty scissors or a knife (Ask an adult to help with this part!)
- PLANTS!

Note that depending on how big an item you're pressing, you can do this with only the strap or only the heavy books for weight. The goal is just to ensure that the plants are pressed down tightly, so wrapping the press with a strap or placing it under a giant pile of books will have the same effect!

STEPS

Depending on the size of the plants you want to press, you can make a small press (think about pressing a four-leaf clover or a bluebell) or a larger one (think about pressing larger tree leaves or a full flower, from blossom to root).
1. Collect your specimens. Make sure you're not taking plants that are endangered or that belong to someone (written permission from the landowner would be good), or that could make your life miserable, like poison ivy!
2. Cut out two or more pieces of cardboard that are bigger than your largest specimen. You can layer

multiple "pressings" in the same press, with cardboard in between each layer.

3. Cut out sheets of newspaper (or regular paper) that match the size of the cardboard.

4. Start building your plant press by layering the items in this order:

    a. Start on a hard, flat surface like a table or the floor.

    b. Lay down your strap, belt, or rope, if using them.

    c. Place one piece of cardboard on top of the strap.

    d. Add one piece of paper.

    e. Lay the plants on the paper. Don't crowd them. You can use a pencil to write on the paper, recording the name of the plant and the date/location and other observations about the habitat where you found it.

    f. Place another piece of paper on top.

    g. Add a piece of cardboard to finish.

    h. Repeat with as many layers as you want, just end with a final piece of cardboard!

    i. Pull the strap tight or pile on the heavy books!

5. Wait about a week to ten days, but remember that many variables can affect how fast your plants dry out. How cold or warm is it? Has it been raining all week? You can lay the plant press near—but not on—a radiator or in the sun to keep things moving. You can also open it and peek in to see how the plants are doing. If the newspaper seems wet, you can carefully change it for dry paper and close it back up.

6. Open the press and C A R E F U L L Y lift up the plants! You can use tweezers or even a spatula if they are really delicate.

7. You did it! You can use these plants in an exhibit, as a guide to local plants, for craft projects, and more!

MAKE A SECCHI DISK

One of the indicators of water quality that scientists study is turbidity, which is related to the cleanliness of the water. Turbid water is murky; that can be caused by a number of factors, from phytoplankton and algae to sediment. While turbidity is not always a problem, too much can signal a water system out of balance, which is bad news for fish and other wildlife. Measuring turbidity is especially helpful when comparing the same body of water over time, or similar bodies of water in the same kind of environment.

Because there is a lot of water in the world, it's hard for scientists to monitor it all. That's where citizen scientists come in! Luckily a tool that Kelly Luis has used in her experiments—the Secchi disk—is one that is inexpensive, easy to use, and can even be made at home. It has not changed much since it was invented, though now there are two different styles: an all-white disk used for measuring ocean turbidity, and a black-and-white one used in freshwater lakes and rivers.

In oceans and lakes around the world, people are using this simple tool to measure the water quality near them and share it with local governments and scientists. For example, a study by the Secchi Disk Foundation invites sailors from all over the world to download an app and share their data from wherever they sail! And in Minnesota and Wisconsin there are lake monitoring programs that invite residents to regularly test their local lakes and report the results.

If you want to test your local waters, or even create an experiment at home or school, it's easy. A store-bought Secchi disk will be a specific size and weight: the ocean disks are all white and 12 inches in diameter, while the freshwater disks are black-and-white and 8 inches. But you can also make one.

To make a Secchi disk you need the following items:
- a flat circular object, like the lid of a paint can
- waterproof paint (white, or black and white, depending)
- a long string (at least 3 feet) (not cotton, which stretches)
- a small weight like a nut and bolt plus a washer

Then, to assemble it:

1. Make a hole in the center of the disk (ask an adult to help use a drill, hammer, or sharp knife).
2. Paint the disk all white, or in alternating quadrants of black and white.
3. Thread the bolt through the hole, add the washer on the underside of the disk, and attach the nut.
4. Attach the string to the bolt.
5. Mark the string at one-foot intervals using a permanent marker.

To use your Secchi disk (in a lake, river, or ocean):

- Slowly lower the Secchi disk into the water, taking care to be on the shaded side of a boat or shore (sunlight interferes with the experiment).
- Stop lowering when the disk is no longer visible.
- Using the marks on the string, record how deep the disk is at this point.
- Repeat the experiment and take an average of the numbers to get a more accurate reading.

To use your Secchi disk in an experiment:

- Get several five-gallon buckets, fill them with water, and add different amounts of dirt, sand, or household substances like flour or cornmeal—anything that won't dissolve!
- Slowly lower the disk into a bucket and note the spot on the string when it disappears, as in the directions above.
- Compare different kinds of sediment and see how it makes a difference in turbidity.

CITIZEN SCIENCE AND MONARCH BUTTERFLIES

Orange-and-black monarch butterflies are beautiful and seemingly fragile creatures that can be found in numerous countries around the world. They're particularly common in North and South America. But the most famous subspecies is a relatively small and endangered group of butterflies that migrates almost three thousand miles from Canada and the northern United States to Mexico. This migration is extraordinary: these tiny, delicate insects travel immense distances, battling weather, predators, starvation, and more to hibernate in the mountains of Mexico. The ones who survive do not make the full return trip in the spring, but instead fly shorter distances to find milkweed and lay eggs. The eggs become caterpillars, then cocoons, then eventually new monarchs that take to the sky and continue north. The cycle continues, with these butterflies laying their eggs and repeating the process. The butterflies that travel north often live only a month or two, but somehow they send their offspring back up to their original homes in Canada. And even more amazingly, some of those offspring have the ability to make the three-thousand-mile journey back to Mexico when autumn returns.

As with so many creatures, monarch butterflies have been impacted by human activity and climate change. They lay their eggs on milkweed, and also depend on other flowers to sustain them over their long journey. But development and the use of pesticides threaten these plants, which in turn makes migration more challenging for monarchs. The number of monarchs that successfully endure this epic migration has gone down in recent years. The International Union for Conservation of Nature has labeled monarch butterflies as endangered.

Organizations from Mexico to Canada have partnered to study these incredible insects, but they need help. And luckily there are a number of ways that people can get involved. The US government has information on several citizen-scientist projects that monitor and protect the migrating monarchs.

But one of the most important—and easiest—ways to get involved in citizen science that helps the monarchs is to plant a garden! Monarch butterflies need rest stops

along their journey, and people are providing waystations that offer them milkweed to lay their eggs on. You don't need a lot of space . . . even one square foot is enough to make a difference. Organizations like the World Wildlife Fund, the National Wildlife Federation, and Monarch Watch have resources on how to plant a monarch-friendly garden. You can then register your waystation so scientists can accurately map how the monarchs can survive on their journey.

SELECTED BIBLIOGRAPHY

#BlackBotanistsWeek. "Rupert Koopman" (resource page).https://blackbotanistsweek. weebly.com/rupert-koopman.html

46 Questions. "Devyani Kumari" (resource page). https://46questions.wordpress.com/2019/07/17/devyani-kumari/

500 Queer Scientists. "Alex Moore" (resource page). https://500queer scientists.com/alex-moore/?ids=

500 Queer Scientists. "Daniel Palacios" (resource page). https://500queer scientists.com/daniel-palacios/

500 Queer Scientists. "Devyani Kumari" (resource page). https://500queer scientists.com/devyani-kumari/

500 Queer Scientists. "Lila Leatherman" (resource page). https://500queer scientists.com/lila-leatherman/

American Chemical Society. "Analytical Chemistry" (resource page). https://www.acs.org/content/acs/en/careers/chemical-sciences/areas/analytical-chemistry.html

American Museum of Natural History. "What is Biodiversity?" (resource page). https://www.amnh.org/research/center-for-biodiversity-conservation/what-is-biodiversity

Andrey Smith, Peter. "Do Seas Make Us Sick? Surfers May Have the Answer." *The New York Times*, April 3, 2017. https://www.nytimes.com/2017/04/03/science/surfers-antibiotic-resistant-bacteria.html

Anita M. S. Marshall, PhD—Geologist. Educator. Photographer. "Home" (resource page). https://anitastonemarshall.com/

ASL-STEM Forum. "About Us" (resource page). https://aslstem.cs.washington.edu/info/about

Association for the Sciences of Limnology and Oceanography. "ASLO honors Caroline M. Solomon with the 2017 Ramón Margalef Award for Excellence in Education" (resource page). May 11, 2018. https://web.archive.org/web/20180511210615/http://aslo.net/awards/2017/solomon.html

Bert, Alison. "On being LGBTQ+ in science - yes it matters, and here's why." Elsevier, July 25, 2019. https://www.elsevier.com/connect/on-being-lgbtq-in-science-yes-it-matters-and-heres-why

Bittel, Jason. "Monarch Butterflies Migrate 3,000 Miles—Here's How." National Geographic, October 17, 2017. https://www.nationalgeographic.com/animals/article/monarch-butterfly-migration

Bortfeld, Victoria. "This 'Green' Space Shouldn't Be So White." Columbia Climate School, August 21, 2020. https://news.climate.columbia.edu/2020/08/21/environmental-sciences-anti-racism/

Botanical Society of South Africa. "Contact" (resource page). https://botanicalsociety.org.za/contact/

Bruckner, Monica Z. "Measuring Lake Turbidity Using a Secchi Disk." Microbial Life Educational Resources. https://serc.carleton.edu/microbelife/research_methods/environ_sampling/turbidity.html

California Department of Fish and Wildlife. "Distinguishing between Coyotes, Wolves and Dogs" (resource page). https://wildlife.ca.gov/Conservation/Mammals/Gray-Wolf/Identification

Calma, Justine. "Black Scientists Call Out Racism in Their Institutions." The Verge, June 11, 2020. https://www.theverge.com/21286924/science-racism-strike-stem-black-lives-matter-protests

Caltech Science Exchange. "How Do We Predict Climate Change?" (resource page). https://scienceexchange.caltech.edu/topics/sustainability/climate-change-predictions

Career Girls. "Oceanographer" (resource page). https://www.careergirls.org/career/oceanographer/

Carrere, Michelle. "To fight climate change, save the whales, some scientists say." Mongabay, March 1, 2020. https://news.mongabay.com/2021/03/to-fight-climate-change-save-the-whales-some-scientists-say/

Chiang-Waren, Xian. "Scientists Call For a 'Justice-Centered Approach to Scientific Research." Audubon, September 14, 2020. https://www.audubon.org/news/scientists-call-justice-centered-approach-scientific-research

CIESIN Thematic Guides. "Satellite Remote Sensing and its Role in Global Change Research" (resource page). http://www.ciesin.org/TG/RS/satremot.html

Connors, Tim. "Using Geologic Maps for Habitat Prediction." https://www.americangeosciences.org/sites/default/files/Environment-capitalreef.pdf

Davidow, Julie. "Seeing himself in the science." University of Washington Magazine, December 2019. https://magazine.washington.edu/feature/ecologist-christopher-schell-sees-himself-in-the-science/

Department: Forestry, Fisheries, and the Environment. "South African history of conservation" (resource page). https://www.dffe.gov.za/projectsprogrammes/peopleparks/southafrican_conservationhistory

Dr. Alex Moore. "Home" (resource page). https://amoorephd.weebly.com/

Dr. Marshall Shepherd. "Current Biography" (resource page). http://www.drmarshallshepherd.com/bio.html

Ecological Society of America. "What Is Ecology?" https://www.esa.org/about/what-does-ecology-have-to-do-with-me/

Editor, WeatherGuys. "Can large cities generate their own weather?" The Weather Guys, July 8, 2019. https://wxguys.ssec.wisc.edu/2019/07/08/cityweather/

Eisenmerger, Ashley. "Ableism 101: What it is, what it looks like, and what we can do to fix it." Access Living, December 12, 2019. https://www.accessliving.org/newsroom/blog/ableism-101/

Environmental Science. "What Is an Environmental Geologist?" (resource page). https://www.environmentalscience.org/career/environmental-geologist

Field Museum. "DIY Plant Pressing" (resource page). July 11, 2017. https://www.fieldmuseum.org/blog/diy-plant-pressing

Flaherty, Colleen. "(More) Bias in Science Hiring." Inside Higher Ed, June 7, 2019. https://www.insidehighered.com/news/2019/06/07/new-study-finds-discrimination-against-women- and-racial-minorities-hiring-sciences

Florida Tech. "What's the Difference between Oceanography and Marine Biology?" June 19, 2018. https://news.fit.edu/archive/difference-between-oceanography-and-marine-biology/

Gabi Serrato Marks. "Home" (resource page). https://www.gabrielaserratomarks.com/

Gallaudet University. "Caroline Solomon" (resource page). https://my.gallaudet.edu/caroline-solomon

Gallaudet University. "Workshop for Emerging Deaf and Hard of Hearing Scientists." May 17–18, 2012. https://www.washington.edu/accesscomputing/sites/default/files/manual-upload/WhitePaper-Final_Gallaudet_Emerging_Sci_2_15_13.pdf

Gartner. "Predictive Modeling" (resource page). https://www.gartner.com/en/information-technology/glossary/predictive-modeling

Goldsmith, Willy. "Stellwagen Bank's Sand Eel Population." On The Water, May 17, 2020. https://www.onthewater.com/stellwagen-bank-sand-lance

Greenfieldboyce, Nell. "'Shrimp on a Treadmill': The Politics of 'Silly' Studies." National Public Radio, August 23, 2011. https://www.npr.org/2011/08/23/139852035/shrimp-on-a-treadmill-the-politics-of-silly-studies

Haeder, Paul. "From Colombia to the Oregon Coast." Discover Our Coast, January 1, 2020. https://www. discoverourcoast.com/oregon-coast-today/columnists/from-colombia-to-the-oregon-coast/article_7c7356ea-236f-11 ea-bc98-3736b3d71462.html

Henshall, Michael. "Conservation is triage | Saving South Africa's flora with botanist-conservationist Rupert Koopman." March 25, 2021. https://www. conservation-careers.com/conservation-jobs-careers-advice/interivews/ conservation-is-triage-conserving-south-africas-flora-with-botanist-conservationist-rupert-koopman/

Hersher, Rebecca. "Why Having Diverse Government Scientists Is Key To Dealing With Climate Change." National Public Radio, April 30, 2021. https://www.npr. org/2021/04/30/981331348/biden-administration-seeks-to-build-trust-and-diversity-among-federal-scientists

How Stuff Works. "What is a computer algorithm?" May 12, 2021. https:// computer.howstuffworks.com/ what-is-a-computer-algorithm.htm\

Hu, Shelia. "Composting 101." National Resources Defense Council, July 20, 2020. https://www.nrdc.org/stories/ composting-101

Illimitable. "Navigating Science with a Disability: Personal Branding and Disclosure." YouTube, October 18, 2020. https://www.youtube.com/watch?v= xdoHg0CnrtY

Institute of the Environment and Sustainability. "Rae Spriggs" (resource page). https://www.ioes.ucla.edu/ person/rae-spriggs/

Johnson, Erika. "Surfing the World for Microbes." UC San Diego News Center, October 6, 2016. https://ucsdnews.ucsd. edu/feature/surfing_the_world_for_ microbes

Katharine Hayhoe. "Home" (resource page). http://www. katharinehayhoe.com/biography/

Keefe, Sean. "NCCS User Spotlight: Dr. Marshall Shepherd." NCCS, https:// www.nccs.nasa.gov/news-events/ nccs-highlights/user-spotlight-shepherd

Kowalski, Kathiann. "An accident didn't stop this geologist from doing field work." Science News for Students, March 3, 2020. https://www.science newsforstudents.org/article/an-accident-didnt-stop-this-geologist-from-doing-field-work

L., Kelly. "Building the Fish Pond with Dad, Hawaii." Smithsonian Institution. https://museum onmainstreet.org/ content/building-fish-pond-dad-hawaii

Lakes Monitoring Program. "How to Use a Secchi Disk" (resource page). https://www.rmbel.info/training/how-to-use-a-secchi-disk/

Lambert, Cash. "The Elements of Cliff Kapono." Freesurf. https://freesurf magazine.com/elements-cliff-kapono/

Langin, Katie. "What does a Scientist Look Like?" Science, March 20, 2018. https://www.science.org/content/article/ what-does-scientist-look-children-are-drawing-women-more-ever

Latham, Byron. "Rupert's seeds for success." MatieMedia, July 10, 2020. https://www.matiemedia.org/ ruperts-seeds-for-success/

Leatherman, Lila. "Science has to do better for its queer, trans, and non-binary scientists." Massive Science, March 31, 2019. https://massivesci.com/ articles/trans-visibility-science-queer-lgtbqia-transgender-inclusion/

Leventry, Amber. "Explaining Nonbinary: How to Talk to Kids About Gender." Parents, March 11, 2020. https://www. parents.com/kids/how-to-talk-to-kids-about-gender/

Lila Leatherman. "Home" (resource page). https://lilaleatherman.com/

Linkedin. "Rupert Koopman" (resource page). https://www. linkedin.com/in/rupert-koopman-89217845/?original Subdomain=za

Locke, Colleen. "UMass Boston Marine Science PhD Student Earns $72K Ford Foundation Fellow- ship." UMass Boston News, July 1, 2018. https://www. umb.edu/news/detail/umass_boston_ marine_science_phd_student_earns_ ford_foundation_fellowship

Manke, Kara. "Systemic racism hurts not just humans, but urban biodiversity." Berkeley News, August 13, 2020. https://news.berkeley.edu/story_jump/ systemic-racism-hurts-not-just-humans-but-urban-biodiversity/

Massive Science. "Gabriela Serrato Marks" (resource page). https:// massivesci.com/people/gabriela-serrato-marks/

McCluney, Courtney L., Katharina Robotham, Serenity Lee, Richard Smith, and Myles Durkee. "The Costs of Code-Switching." Harvard Business Review, November 15, 2019. https://hbr. org/2019/11/the-costs-of-codeswitching

Meteorology (resource page). https:// www.nationalgeographic.org/ encyclopedia/meteorology/Million Stem. "Devyani Singh" (resource page). https://www.1mwis.com/profiles/ Devyani-Singh

Minas, Shant. "How Geology Might Help Us Understand Climate Change." Applied Earth Sciences, October 30, 2018. https://www.aessoil.com/how-geology-might-help-us-understand-climate-change/

Monarch Watch. "Monarch Waystation Program" (resource page). https://www. monarchwatch.org/waystations/

Musavengane, Regis, and Llewellyn Leonard. "When Race and Social Equity Matters in Post-apartheid South Africa." Conservation & Society. https://www.jstor.org/stable/26611740?seq=1#metadata_info_tab_contents

NASA Climate Kids. "How Do We Predict Future Climate?" (resource page). https://climatekids.nasa.gov/climate-model/

National Centers for Environmental Information. "What is Paleoclimatology?" (resource page). https://www.ncdc.noaa.gov/news/what-paleoclimatology

National Geographic. "Atmosphere" (resource page). https://www.nationalgeographic.org/topics/resource-library-atmosphere/?q=&page=1&per_page=25&page=1&per_page=25

National Geographic. "Paleontology" (resource page). https://www.nationalgeographic.org/encyclopedia/paleontology/ 6th-grade/

National Invasive Species Information Center. "What are Invasive Species?" (resource page). https://www.invasivespeciesinfo.gov/what-are-invasive-species

National Ocean Service. "What does an oceanographer do?" (resource page). https://oceanservice.noaa.gov/facts/oceanographer.html

National Oceanic and Atmospheric Administration. "What is a mangrove forest?" https://oceanservice.noaa.gov/facts/mangroves.html

Native Nations Climate Adaptation Program. "Valerie Small (Apsáalooke'-Crow)" (resource page). https://www.nncap.arizona.edu/node/117

North American Nature. "What Is The Difference Between a Coyote and a Wolf?" (resource page). https://northamericannature.com/what-is-the-difference-between-a-coyote-and-a-wolf/

Oregon State University. "Daniel M. Palacios" (resource page). https://mmi.oregonstate.edu/people/daniel-m-palacios

Ottesen, KK. "An evangelical scientist on reconciling her religion and the realities of climate change." The Washington Post, March 2, 2021. https://www.washingtonpost.com/lifestyle/magazine/an-evangelical-scientist-on-reconciling-her-religion-and-the-realities-of-climate-change/2021/02/26/f757d1c2-40b7-11eb-8bc0-ae155bee4aff_story.html

Pacific Northwest Research Station. "Monitoring Forests From Space: Quantifying Forest Change by Using Satellite Data." January 2007. https://www.fs.fed.us/pnw/sciencef/scifi89.pdf

Parkin, Benjamin. "Crisis in the Himalayas: Climate change and unstable development." Los Angeles Times, March 22, 2021. https://www.latimes.com/world-nation/story/2021-03-22/himalayas-climate-change-unsustainable-development

Pew Charitable Trusts. "Do Health Impact Assessments Help Promote Equity Over the Long Term?" (resource page). November 11, 2020. https://www.pewtrusts.org/en/research-and-analysis/reports/2020/11/do-health-impact-assessments-help-promote-equity-over-the-long-term

Poulsen, Zoë. "What is Fynbos?" Notes from a Cape Town Botanist. https://www.capetownbotanist.com/what-is-fynbos/

Prodanovich, Todd. "No Man Is an Island: The Cliff Kapono Profile." Surfer, April 25, 2019. https://www.surfer.com/features/no-man-is-an-island-the-cliff-kapono-profile/

Pyzel, Mara. "Surfer & Scientist Cliff Kapono Celebrates Earth Day Every Day." Freesurf. https://freesurfmagazine.com/surfer-scientist-cliff-kapono-celebrates-earth-day-every-day/

Quivira Coalition. "Valerie Small" (resource page). https://quiviracoalition.org/valerie-small/

Rady Children's Hospital. "Raenita Spriggs" (resource page). https://sdhealthscholars.org/team-member/raenita-spriggs/

Rocío Paola Caballero-Gill, Ph. D. "About" (resource page). https://rociocaballerogill.weebly.com/about.html

Rocky the Rock Ripper. "At-Home Science Experiments: Secchi Disk." On the Waterfront, March 31, 2020. http://blog.waterfrontoronto.ca/nbe/portal/wt/home/blog-home/posts/at-home-experiments-secchi-disc

Ruf, Jessica. "Why Environmental Studies is Among the Least Diverse Fields in STEM." Diverse Education, February 16, 2020. https://diverseeducation.com/article/166456/

SACNAS. "The Mountains Are Calling & We Can All Go! Accessibility, Inclusion & Innovation in the Geosciences." YouTube, June 26, 2020. https://www.youtube.com/watch?v=dpREXIhyK7A

Sanchez, Claudio. "Biology Professor's Calling: Teach Deaf Students They Can Do Anything." National Public Radio, May 20, 2015. https://www.npr.org/sections/ed/2015/05/20/406148448/in-the-classroom-and-on-the-river-modeling-success-in-science

Sanibel Sea School. "Day 6: Plant Pressing" (resource page). https://www.sanibelseaschool.org/naturenearyoulessons/2020/3/31/day-4introduction-to-xylem-and-phloem-le6cd-d9dja

Schwartz, John. "Katharine Hayhoe, a Climate Explainer Who Stays Above the Storm." The New York Times, October 10, 2016. https://www.nytimes.com/2016/10/11/science/katharine-hayhoe-climate- change-science.html

Secchi Disk. "Secchi Disk" (resource page). http://www.secchidisk.org/

Skelton, Renee, and Vernice Miller. "The Environmental Justice Movement." National Resources Defense Council, March 17, 2016. https://www.nrdc.org/stories/environmental-justice-movement

South African National Biodiversity Institute. "Custodians of Rare and Endangered Wildflowers (Crew) Programme" (resource page). https://www.sanbi.org/biodiversity/building-knowledge/biodiversity- monitoring-assessment/custodians -of-rare-and-endangered-wildflowers-crew-programme/

Stone, Maddie. "Weathercasters are talking about climate change — and how we can solve it." Grist, January 28, 2020. https://grist.org/climate/weathercasters-are-talking-about-climate-change-and-how-we-can-solve-it/

Stone, Madeleine. "How much is a whale worth?" National Geographic, September 24, 2019. https://www.nationalgeographic.com/environment/article/how-much-is-a-whale-worth

The Conversation. "We mapped green spaces in South Africa and found a legacy of apartheid." https://theconversation.com/we-mapped-green-spaces-in-south-africa-and-found-a-legacy-of-apartheid-143036

The National Institute of Social Sciences. "What is 'Social Science?'" (resource page). https://www.socialsciencesinstitute.org/what-is-social-science

Ullman, Dave. "Geologic Perspectives on Climate Change." Northland College, September 7, 2017. https://www.northland.edu/news/geologic-perspectives-climate-change/

Understanding Science. "Who pays for science?" (resource page). https://undsci.berkeley.edu/article/who_pays

United State Environmental Protection Agency. "Environmental Justice." https://www.epa.gov/environmental justice

United States Department of Agriculture. "Remote Sensing in Forest Health Protection" (resource page). https://www.fs.fed.us/foresthealth/technology/pdfs/RemoteSensingForestHealth00_03.pdf

United States Environmental Protection Agency. "Composting At Home" (resource page). https://www.epa.gov/recycle/composting-home

United States Environmental Protection Agency. "Cyanobacteria Assessment Network Application (CyAN app)" (resource page). https://www.epa.gov/water-research/cyanobacteria-assessment-network-mobile-application-cyan-app

United States Environmental Protection Agency. "Environmental Justice" (resource page).https://www.epa.gov/environmentaljustice

United States Geological Survey. "Satellite Data Shows Value in Monitoring Deforestation Forest Degradation." April 2, 2021. https://www.usgs.gov/center-news/satellite-data-shows-value-monitoring-deforestation-forest-degradation?qt-news_science_products=1#qt-news_science_products

United States Geological Survey. "What is remote sensing and what is it used for?" (resource page). https://www.usgs.gov/faqs/ what-remote-sensing-and-what-it-used?qt-news_science_products=0#qt-news_science_products

Weinberger, Hannah. "UW research shows racism and redlining hurt local wildlife, too." Crosscut, August 20, 2020. https://crosscut.com/environment/ 2020/08/uw-research-shows-racism-and-redlining-hurt-local-wildlife-too

Wetlands International and The Nature Conservatory. "Mangroves for coastal defence" (resource page). https://www.nature.org/media/oceansandcoasts/mangroves-for-coastal-defence.pdf

Whales & Climate Research Program. "Welcome to Whales & Climate Research Program" (resource page). https://whalesandclimate.org/

What Is an Urban Heat Island? (resource page). https://climatekids.nasa.gov/heat-islands/

World Wildlife Fund. "Eastern Himalaya" (resource page). https://wwf.panda.org/discover/knowledge_hub/where_we_work/eastern_himalaya/solutions2/climate_change_solutions/

Yurkiewicz, Ilana. "Study shows gender bias in science is real. Here's why it matters." Scientific American, September 23, 2012. https://blogs.scientificamerican.com/unofficial-prognosis/study-shows-gender-bias-in-science-is-real-heres-why-it-matters/

Zak, Dan. "One of America's top climate scientists is an evangelical Christian. She's on a mission to persuade skeptics." The Washington Post, July 15, 2019. https://www.washingtonpost.com/lifestyle/style/one-of-americas- top-climate-scientists-is-an-evangelical-christian-shes-on-a-mission-to-convert-skeptics/2019/07/12/9018094c-8d2a-11e9-adf3-f70f78c156e8_story.html

SOURCE NOTES

All quotations are from interviews with the author, except as noted below.

Alex Moore
p. 19 "trying to lead . . . up to you."
Bert, Alison. "On being LGBTQ+ in science – yes it matters, and here's why." Elsevier website, July 25, 2019. https://www.elsevier.com/connect/on-being-lgbtq-in-science-yes-it-matters-and-heres-why

p. 20 "When I . . . my identities." Ibid.

Lila Leatherman
p. 65 "Science . . . of a society."
Leatherman, Lila. "Science has to do better for its queer, trans, and nonbinary scientists." Massive Science website, March 31, 2019. https://massivesci.com/articles/trans-visibility-science-queer-lgtbqia-transgender-inclusion/

p. 70 "Feeling unable . . . you're experiencing." Ibid.

Marshall Shepherd
p. 77 "first introduced . . . scientific method." "NCCS User Spotlight: Dr. Marshall Shepherd." NASA Center for Climate Simulation website. https://www.nccs.nasa.gov/news-events/nccs-highlights/user-spotlight-shepherd

Daniel Palacios
p. 94 "Today the . . . human consumption." Haeder, Paul. "From Colombia to the Oregon Coast."

Discover Our Coast, Jan 1, 2020. https://www.discoverourcoast.com/oregon-coast-today/columnists/from-colombia-to-the-oregon-coast/article_7c7356ea-236f-11ea-bc98-3736b3d71462.html

p. 98 "Because homophobia . . . LGBTQ people." Profile on 500 Queer Scientists website. https://500queerscientists.com/daniel-palacios

Chris Schell
p. 133 "I hope . . . might be incomplete." Manke, Kara. "Systemic racism hurts not just humans, but urban biodiversity." Berkeley News, August 13, 2020. https://news.berkeley.edu/story_jump/systemic-racism-hurts-not-just-humans-but-urban-biodiversity/

Caroline Solomon
p. 162 "I became . . . about it."
Interview posted on USA Deaf Sports Federation website, May 20, 2020. http://usdeafsports.org/news/usadsf-spotlight-with-noah-us-deaflympian-carrie-miller-solomon

p. 164 "One thing . . . just everywhere." Interview on National Public Radio's Morning Edition, May 20, 2015. https://www.npr.org/sections/ed/2015/05/20/406148448/in-the-classroom-and-on-the-river-modeling-success-in-science

p. 168 "Growing up . . . side of things." Ibid.

215

INTERVIEWS/EMAILS WITH SCIENTISTS
(all video/phone interviews unless noted)

Leatherman, Lila. December 2, 2020.
Moore, Alex. December 7, 2020.
Palacios, Daniel. December 7, 2020.
Marshall, Anita. December 14, 2020.
Shepherd, Marshall. December 14, 2020.
Singh, Devyani. December 18, 2020.
Small, Valerie. December 21, 2020.
Solomon, Caroline. December 28, 2020. (via email)
Spriggs, Rae. January 8, 2021.
Hayhoe, Katharine. January 13, 2021.
Schell, Chris. January 21, 2021.
Serrato Marks, Gabriela. January 21, 2021.
Koopman, Rupert. January 26, 2021.
Kapono, Cliff. February 10, 2021.
Caballero-Gill, Rocío Paola. February 22, 2021.
Luis, Kelly. February 22, 2021.
Luis, Kelly. February 26, 2021.

PHOTO CREDITS

ACKNOWLEDGMENTS

There's a famous (or maybe infamous) expression that's often used as a metaphor for writing a book. It goes, "How do you eat an elephant?"

And the answer is: "Bite by bite."

Bite.

By.

Bite.

(NOTE: I do not eat elephants, nor do I support or endorse the eating of elephants. This is a metaphor, not an instruction manual!)

The point, in case you're now thoroughly confused and slightly nauseated, not to mention concerned about my eating habits, is this: When dealing with a giant and seemingly impossible task, all you can do is break it down into manageable pieces and tackle it step by step by step.

Breaking the Mold is my seventh published book for young readers, and the hardest one I've ever written. Nonfiction books are a whole different kettle of fish (kettle of elephants?!), and the learning curve was a vertical climb. I had no idea what I was getting myself into, and once I *did* realize, I simply had to take it step by step. I would never had made it without the help and support of so many incredible people.

First and foremost, huge thanks to friends like Erika Spanger-Siegfried at Union of Concerned Scientists, and Dr. Cameron Wake, Josephine A. Lamprey Professor in Climate and Sustainability at the University of New Hampshire Sustainability Institute, who talked to me about the big-picture elements of climate change and why we need to get more kids of all backgrounds aware and involved. When this idea was just a spark, they urged me to write it and bring these voices to life.

I also want to throw a huge virtual thank-you party for the nonfiction WOMGeniuses of my middle grade writing world, and in particular: Sarah Albee, Uma Krishnaswami, Kate Messner, and Kekla Magoon, who shared so much brilliance that I practically needed sunglasses to listen to them. These folks have written some of the funniest, most exciting, most compelling, most entertaining, and most delightful nonfiction books you'll ever read. Their subjects range from poisons to pythons, from the the Black Panther Party to Gandhi. I mean really...go read them all.

The folks at Holiday House, especially my amazing editor Della Farrell, have turned my good idea into a much better book. I am so grateful for Della's wisdom and kindness and steady editorial hand as I flailed around learning how to write this. And another huge thanks goes out to Kerry Martin and Ed Miller, who took 200 pages of text and turned them into something so visually beautiful and exciting I can hardly believe it. And of course I am deeply grateful to George Newman and Hayley Jozwiak for copyediting and proofreading and helping correct my mistakes. Any that remain are, of course, my own responsibility.

Additional thanks also to the invaluable Gabriela Murray, who helped me wrangle all the loose ends of endnotes and quotations into neat and tidy spreadsheets—if she had not jumped in, I would probably still be trying get them all finished. Thank you, Gabi! Then there are the ride-or-die gang: my beloveds, Patrick, Noah, and Izzy; super-agent Marietta Zacker and the Gallt & Zacker Agency crew; the amazing MoB . . . maybe I could do this without you, but I hope I never have to try.

And finally, my profound and humble and grateful thanks to all of the scientists past, present, and future, in this book and beyond, who are breaking the mold. You are pushing back against tough barriers. You are taking a path riddled with roots and hidden thorns, and fighting headwinds that are not always apparent to your colleagues. You are reminding us all, again and again, that this world belongs to all of us, and that we all have work to do. You deserve to be celebrated. You are amazing, and the world is a better place because you're in it.

INDEX

Bold numbers indicate definitions.

African Design

by the same author

*

AFRICAN TAPESTRY
(Autobiographical)

*

CLASSICAL AFRICAN SCULPTURE

*

AFRICAN ARTS AND CRAFTS
(*Longmans*, 1937)

ART TEACHING IN AFRICAN SCHOOLS
(*Longmans*, 1952)

*

With K. P. Wachmann

TRIBAL CRAFTS OF UGANDA
(*Oxford University Press*, 1953)

RAFFIA PILE CLOTH

Bushongo Tribe   *Belgian Congo*

AFRICAN DESIGN

Margaret Trowell

FREDERICK A. PRAEGER, *PUBLISHERS*

New York

BOOKS THAT MATTER

Published in The United States of America in 1960
by Frederick A. Praeger, Inc., Publishers
64 University Place, New York 3, N.Y.
All rights reserved
Printed in Great Britain

Library of Congress catalog card number 60–11832

Contents

Illustrations

Raffia Pile Cloth   *frontispiece*

(The Plates are at the end of the book)

There is a map facing page 13

Acknowledgments

Thanks are due to the following museums for permission to reproduce photographs of specimens from their collections:

Cambridge, Museum of Archaeology and Ethnology: XLb.

Dar es Salaam, King George V Museum: XLa.

London, British Museum: XVc, XVIa, d, XVIIa, b, XVIII, XIXd, XXa, b, XXI, XXIIa, XXVII, XXVIIIa, b, XXIXa, b, XXXIIId, XXXIVa, XXXVIa, b, d, XXXVIIIa, b, XXXIXa, b, XLc, d, XLIc, XLIIa, b, c, d, XLIII, XLVIb, La, b, LIa, c, LIIa, b, d, LIIIa, b, c, LVa, LVIIIa, b, LIXc, d, LXIa, b, LXIIa, b, LXIIIb, LXIVa, b, LXVa, b, c, d, LXVIa, b, c, d, LXVIIa, b, c, d, LXVIIIa, b, LXIXa, LXXd, LXXIIb, c, LXXIIIa, c, LXXIVa, LXXV, LXXVIc, d.

Manchester, The University Museum: XXXIIIc, XLIa, b, XLVIa, LXIIIa.

New York, The American Museum of Natural History: LIVc, d, LVI, LXXb.

Oxford, Pitt Rivers Museum: XVIb, XVIIc, d, XIXa, b, c, XXIIb, XXIIIa, b.

Paris, Musée de l'Homme: XXV, XXIXc, XXXIa, b, XXXIIa, b, XXXIIIb, XLVIIa, b, XLVIIIa, b, c, d, XLIXa, b, LVb, LX, LXXIIa.

Tervuren, Musée Royal du Congo Belge: Xa, b, XIa, b, c, XVb, XVIc, XXIVb, XXVIa, b, XXIXd, XXXIIIa, XXXIVb, XXXVa, d, XXXVIc, XXXVIIa, b, XXXIXc, XLIVb, XLVb, c, LIIc, LIVa, b, LVc, LIXa, b, LXXc, LXXIb, c, d, LXXIIIb, d, LXXIVb, c, d, LXXVIa, b.

Uganda Museum: XIIa, b, XXXVb, c, LIb, LXXa, LXXIa.

Zanzibar, Peace Memorial Museum: XIVa, b.

And to the following writers, publishers and photographers for other photographs:

Plate XLVa is taken from *Afrika, Handbuch de angewandten Volkerkunde* by Univ. Prof. Dr Hugo A. Bernatzik, by kind permission of the author's widow.

Dr P. Bohannan and the Editor of *Man*: XLVd.

The Editor of *Brousse*: IIIb, IVc, IXa.

9

ACKNOWLEDGEMENTS

Mr H. Cory and Faber and Faber: IIa.

Mr E. H. Duckworth: VI, VII, XXXa, XLIVa.

Mr E. H. Duckworth and the Editor of *Nigeria*: Ia, b, VIIIa, XVa, XXXb, LXIXb.

Mr Gomez: LVIIIc.

Mrs Meyerowitz and Faber and Faber: XXIVa, LXVIIc, d.

Mr P. Moeschlin: LVII.

The Editor of *Nigeria*: IIb.

L'Institut d'Ethnologie, Paris: Va, b.

Mr J. Redinha and the Companhia de Diamantes de Angola: IIIa.

Uganda Information Department: VIIIb.

As is usual in a study of this kind individual thanks are due to far more people than can be mentioned here, among them many members of the staff of the museums listed above. More especially would I thank Miss Bennet-Clark and Mr William Fagg of the British Museum, and Miss Blackwood of the Pitt Rivers Museum; Mrs Ehrlich, Mrs French and Mrs Tattersall for typing the manuscript; the Musée Royal du Congo Belge for a magnificent gift of photographs of specimens from their collections, many of which are reproduced here; the Council of Makerere College for a research grant and a term of study leave; and last but not least my publishers for the detailed attention and patience which they always give to their publications.

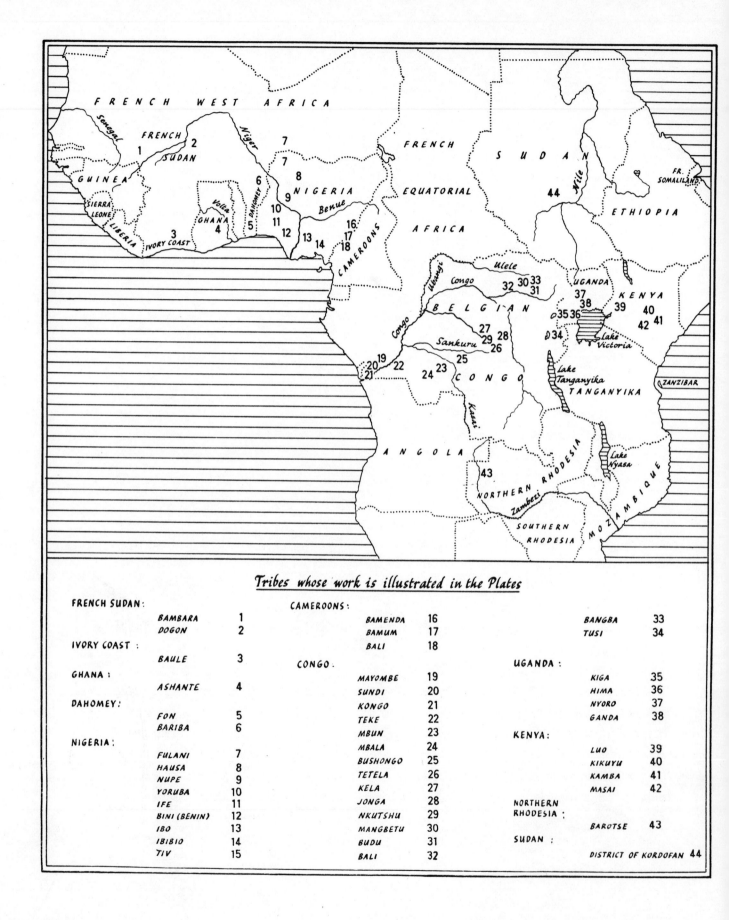

Tribes whose work is illustrated in the Plates

FRENCH SUDAN:

BAMBARA	1
DOGON	2

IVORY COAST:

BAULE	3

GHANA:

ASHANTE	4

DAHOMEY:

FON	5
BARIBA	6

NIGERIA:

FULANI	7
HAUSA	8
NUPE	9
YORUBA	10
IFE	11
BINI (BENIN)	12
IBO	13
IBIBIO	14
TIV	15

CAMEROONS:

BAMENDA	16
BAMUM	17
BALI	18

CONGO:

MAYOMBE	19
SUNDI	20
KONGO	21
TEKE	22
MBUN	23
MBALA	24
BUSHONGO	25
TETELA	26
KELA	27
JONGA	28
NKUTSHU	29
MANGBETU	30
BUDU	31
BALI	32

BANGBA	33
TUSI	34

UGANDA:

KIGA	35
HIMA	36
NYORO	37
GANDA	38

KENYA:

LUO	39
KIKUYU	40
KAMBA	41
MASAI	42

NORTHERN RHODESIA:

BAROTSE	43

SUDAN:

DISTRICT OF KORDOFAN	44

CHAPTER I

The Craftsman's Aim

The 'fine' arts of painting and sculpture have always been considered to be of much greater aesthetic value than the 'applied' arts, by which we mean the various branches of design and ornament applied to the making of our material possessions. In one sense it is good that this should be so, good that even in a material age man instinctively rates higher an activity directed towards a purely aesthetic end rather than towards a purely utilitarian one. But in another sense it is the result of shallow thinking, for if we take the trouble to define our terms and study recognized masterpieces of both fine and applied art we find that both stem from the same roots; man's personal desire to create things of beauty, his need to use his products in the service of his community, and his wish to link himself with the spiritual power or powers behind the visible world through the use or offering of his own small creation.

These intangible urges are obvious in the fine arts but they also exist in the applied ones. We speak of design in a craft, and by that we mean the whole process of planning the shape of an object and its construction in a way which is not only functionally satisfactory but also pleasing to the eye and touch, and this satisfaction is both utilitarian and aesthetic; then we include ornament in design and this element is not in the least essential to the utility of the object but owes its existence solely to the desire of the craftsman and the user for something over and above efficiency and comfort. As for the other considerations which we have suggested above, that is the use of the artist's and craftsman's products for the service of the community and as a link with spiritual powers, here the applied and the so-called fine arts are analogous, both are often commissioned for the service of the social power of the community to enhance its prestige and satisfy its desire for things of beauty, and both may through actual representation or symbolism play an important part in its spiritual and psychological life.

On the other hand the more technical qualities of fitness for purpose and respect for the possibilities and limitations of materials, tools and techniques which are obviously essential for a satisfying design, are really equally important in a painting or a piece of sculpture, though this may not be so obvious to the layman. The difference, then, between a work of fine art and a design for a craft is not absolute but is rather one of emphasis and degree. In a primitive or pre-industrial society even this difference is far less than in a more sophisticated one, for here life is an integrated whole and in both the sculptured ancestor figure and the decorated calabash the practical and the more intangible creative qualities are fairly evenly balanced, for the sculpture is made for the very practical purpose of harnessing spiritual power for material ends, while the decorated calabash has a recognized symbolical significance as well as a material use. The distinction between fine and applied art would puzzle the primitive artist.

We are sometimes apt to designate the craftsmanship of primitive peoples as 'folk art', a term which we use in Western culture for the artistic creations of peasant folk for their own ends. This is somewhat erroneous for although many objects are made by the primitive craftsman for his own household use, yet others such as textiles, food vessels, weapons and wood carving may be created by him for the use of the great chiefs or for special ceremonial purposes within a social system which has, in its own fashion, some resemblance to that which produced the hangings and the furnishings for the Versailles of Louis XIV, and its products will bear the stamp of sophistication of a court art.

The physically controlling factor in the development of all forms of art would seem to be architecture, which is itself largely controlled by the materials available for building. Architecture provides scope and security for all other works of art. Large sculpture, paintings and objects of decorative design are required as a part of the building itself, while smaller portable ones can only be preserved from damage by weather or insects when they are properly housed. The pastoral nomad may have as good an aesthetic sense as his more sedentary brother, indeed in Africa he seems to have that sense more strongly developed than many of the agricultural tribes, but his way of life will give him no encouragement to make large pieces of sculpture or paintings, and even in the decorative crafts he will restrict his skills to weaving, basketry, decorated calabashes and articles of personal adornment rather than heavy, bulky or fragile objects. Among the more sedentary tribes it is those people whose religious customs or social pattern of kingship have led them to develop some form of architecture who have also developed the decorative arts, so that we turn to the West Coast Kingdoms or such people as the Bushongo of the Congo for our best examples of the crafts.

We cannot in fact fit the art of primitive peoples into the exact categories which we use for that of our own society; nevertheless we find that these people, like ourselves, have the fundamental urge to create things of beauty, and through the study of art we must attempt to understand the expression of this common urge in terms of the material and social background of other races and ages. Tradition, fashion, taste and style are all linked to background, as we realize at once when we consider our own art history. We think of the solidity and strength of Romanesque architecture and design which was essential for the life of its time; then of the changes which came through growing technical knowledge and the control of materials as well as the altered social conditions of the later days of Gothic architecture; and, the most striking change of all, of our own day where the combination of vastly increased control of material on the one hand and the lack of servants on the other has created a completely new style of domestic architecture and interior design.

Styles change but the urge is the same. The ladies of Europe in the Middle Ages passed their hours in embroidering wonderful tapestries and hangings for the castle walls. Those of the more cultured tribes of Africa may not have the same wealth of coloured silks at their disposal, but with raffia and vegetable dyes they weave and embroider the finest patterned cloths or make the most delicate basketry. Again in Medieval Europe the monk ornamented the margins of his text with delightful bab-wyneries and the carver who was employed in building the cathedral decorated the misericords and corbels and waterspouts with the same fantastic types of beasts or with whimsical genre scenes and figures; much of the ivory carving of Benin and wooden panelling of the Yoruba suggests the same approach in the treatment of fabulous reptiles and scenes from everyday life. Heraldry in Europe is paralleled by the appliquéd cloths of Dahomey and Ghana in which the power and might of the chiefs are often expressed by symbolic forms or scenes; while the ceremonial staves and axes and paddles of many African tribes have a refinement and dignity of design which compare favourably with the regalia of more technically advanced peoples in spite of their limited choice of materials. If we are to appreciate the art of a pre-industrial society such as Africa we must study it as a whole, considering the decoration of pots and textiles and wooden vessels as closely as we would the carving of figures and masks. These examples of decorative design may not make the same violent impact on our minds as did African carving on those of Western artists a decade or two ago, but they may well bring fresh vitality to our approach to the purpose and the pleasure of pattern design.

We began by defining design as the whole planning of the shape and construction of an object, using the word in its widest sense, and we should remember that even when the term is used in the more limited field of decoration or ornament it carries with it this wider connotation. By design used in the more restricted sense we mean the decoration of an object such as a wall, panel, pot, basket, vessel or cloth with a pattern in which one or more motifs are repeated in a rhythmical manner; or, if they are not repeated, are fitted together to make a balanced whole within the decorated form. Such patterns are often abstract or geometrical, but representational forms of human, animal or floral types or material objects may be used provided that their treatment is decorative rather than photographic in intention. Such formalized representation is often symbolic.

A study of pattern may be approached in a number of different ways. There is first the technological approach. Working along this line we must study our specimens from the point of view of the possibilities and limitations of the materials and tools which have been used. If the basic material is something hard like stone or metal or wood, or bone or ivory, we shall find it has been carved or engraved, or it may have had other material such as thin sheets of metal appliquéd on to its surface in places, or a contrasting metal or wood or enamel inset, or the whole or part of the object may have been coated with paint. Metal too may be cast, in which case it will be subjected to entirely different treatment.

Clay may be ornamented in a number of different ways, for while it is still soft it may be moulded, or when it is semi-hard it may be impressed or stamped, or engraved or scratched, or later still it may be decorated with a slip of clay of a different colour, or burnished or painted with a coloured glaze.

Fibres again have their appropriate treatment. They may have the pattern woven in the basic structure, or it may be printed on by several different methods, or they may be embroidered with different types of fibre or ornamented by the application of different materials altogether. Each material, whether it be used as the base or as part of the decoration, will give the designer scope for different forms of treatment, and will also have its natural limitations.

Then we must know what kind of tools the artist has used, for these again have their natural possibilities and limitations. An African adze produces a very different surface texture from European carving tools, both have their own beauty and they should never be confused. Again the slight irregularity of handblock printing gives a very different effect from the slick finish of machine printing; both have their place, and a strained effort after mechanical perfection on the part of the handcraftsman is as unpleasant and false as the imitation of irregular handprinting by a machine. Apart from the actual impression of the tool which is used on the material an apparently small variation in technique will open up a whole new range of possible pattern forms.

THE CRAFTSMAN'S AIM

From this consideration of materials, tools and techniques we may be tempted to theorize upon matters which are not strictly speaking the concern of the designer. For directly we begin to study the material culture of a people, whether it be the artefacts which they use in everyday life, or the way in which they decorate them, or the non-utilitarian objects which they make for religious or aesthetic purposes, we are at once struck by the pointers which they give us concerning the relationship between one tribe and another. Such and such are the musical instruments of a certain number of tribes; such is the technique used in the basket making of another group, or particular pattern motifs keep cropping up in the ornament of a third set. While it is dangerous to jump to conclusions on flimsy evidence it is at least suggestive that here we have a clue to racial or tribal affinities or a common geographical past.

Then, when we come to study the pattern motifs of these groups more closely in the desire to clarify their apparent similarities and differences, we realize that the whole field of stylization and symbolism is opened before us, and we must study the way in which representational forms, symbols and geometrical pattern are worked out independently from technical considerations. In an inclusive account of the decorative art of Africa both the ethnologist's and the craftsman's approach are equally necessary. A detailed study of the use of the human form in design with all the variations and degrees of stylization which can be found would fill a volume in itself, and the same may be said of any other motif from the whole vocabulary of symbolism. Here little has been attempted beyond discussing the fundamental possibilities and limitations set by technique in applied design, and following it with a description of some pattern forms which seem to be of outstanding interest. The use of human and animal motifs has been illustrated from the work of parts of the West Coast where it seems to be most highly developed; while as the field of geometrical pattern seems to be richest in the Congo our discussion of this type of design has been centred there. The approach is frankly that of an artist rather than of an ethnologist, and makes no pretence of being a detailed survey of the subject. But it has been written in the belief that such an outline would be of interest to the student of art history and to the practical designer, be he African or European.

It has been difficult to decide on the geographical area which should be included in such a study, or to choose a satisfactory title for that area. It would be too big a task to cover the art of the Islamic peoples of North Africa or that of Abyssinia, and it has therefore seemed best to confine the study to what is often termed Africa South of the Sahara, using the term Negro Africa in a very loose sense to cover any African peoples found in that area, and acknowledging definite Moslem influence in the art of such districts as Northern Nigeria and the Coast of East Africa.

CHAPTER II

Wall Decoration

Rocks and the walls of caves and, as time went on, those of mud huts have presented an attractive surface for decoration since very early times; indeed the earliest two-dimensional art which still exists was carried out on rock walls for, after all, they were the only flat surfaces available. They were painted in earth colours together with soot and perhaps vegetable dyes, so that the colour scheme was restricted to blacks and browns, red and yellow ochres, greys and whites. These rock surfaces had no definitely closed boundary or form, and later the circular wall of a mud hut had little more, so that the sense of necessity for regular pattern was not at first developed, and even today many cave and wall paintings exist in which decorative treatment *per se* is completely lacking. As in all forms of decorative art two trends, the one towards naturalism and the other towards stylization and symbolism, were always at work. On the one hand we find lively and realistic scenes of hunting and battle depicted, on the other we notice forms being reduced to an accepted pictorial shorthand which often seems to have almost no connection with the objects after which the symbols are named. Both representational drawings and symbols are often dotted about the wall surface with no apparent attempt at arrangement. Various techniques are used, the work may be painted, or modelled in low relief in clay, or first modelled and then painted.

It would seem that the initial purpose of both the naturalistic and the symbolic forms used in this way was purely utilitarian. Both had some magico-religious function, and their value for the owner lay in this alone. But man is essentially a creator, and he naturally tends to delight in his creation and to strive after beauty for its own sake. Decoration brings its own discipline, and the artist begins unconsciously to arrange both the representational and symbolic forms within definite panels; he combines the abstract with the realistic to provide borders, or to fill up spaces in the design, and where he uses abstract forms alone he arranges them in an orderly and decorative fashion with due regard to the space which they are intended to fill.

Naturalistic Mural Painting
(Plates I, II)

It is in Nigeria that we find the most developed representational mural painting, although it is possible that equally good work may exist in other parts of Africa without having been photographed or recorded. But certainly the large-scale paintings on the walls of Mbari houses, Court houses and chiefs' dwellings in Nigeria are most remarkable for their vitality. Here great stretches of wall are decorated with brightly coloured paintings of scenes from everyday life, broken up by borders and panels of pattern, the works combining realism and decorative quality in a way which is exactly suited to mural painting. The wall paintings by the snake charmers in Tanganyika and those recorded from hut walls in Angola also show the qualities of mural painting in a well-developed form, many of those from Angola most successfully combining stylized human forms with abstract or symbolic pattern.

Abstract Design in Mural Painting
(Plates III, IV)

In every area where mural painting is practised representational art merges by easy stages into completely abstract pattern or symbolism. A most extraordinarily interesting series of wall paintings have been recorded by Scohy[1] from the village of Ekibondo in the Uele region of the north-east Congo. Here, through the encouragement of an enlightened District Commissioner, a traditional craft of wall painting has been revived and the external hut walls are covered with rich geometrical pattern and panels containing men, animals, fish and other objects. The patterning of the walls is most striking in its boldness and spontaneity.

[1] Scohy, 'Ekibondo'. *Brousse.* 1951, pp. 1–2.

WALL DECORATION

NATURALISTIC LOW RELIEF PANELS IN CLAY, WOOD AND METAL
(*Plates V, LVII, LVIII, LXVIII, LXIX*)

Low relief panels in clay, wood and metal, with subjects treated in a naturalistic way, are found in many parts of West Africa. Whether these should be considered as applied design is difficult to decide for although some, through their stylized treatment and emphasis on arrangement, must definitely be regarded in this light, others must be thought of more in sculptural terms. Schmalenbach[1] has pointed out that the sculpture of the African is very largely limited to individual figures. In the more primitive reliefs, too, figures are introduced quite without relationship to one another either by form or pattern, or by any feeling of tension or articulation. As we would say from the angle of this study, there is no sense of design. The first step towards planned design in low relief carving would seem to be shown in the carved door panels of such tribes as the Dogon and Senufo of the Western Sudan, where figures are ranged side by side, or one above the other, in almost military formation; the convention of frontality being adhered to as rigidly as in Byzantine painting. It is only in the work of the coastal kingdoms that more freedom is achieved, and the artist begins to compose his design with a definite sense of rhythm and connection between his units of pattern.

Clay panels in low relief were used to decorate the palace walls of kings and those of the houses of chiefs and juju priests in Dahomey and Nigeria, but the early European explorers with their rigid conventions of representational art found them of little interest. Lander, speaking of some he had seen in Nigeria, says 'the execution, as might be supposed, is rude and contemptible'. Some of the panels from the walls of the royal palaces of Dahomey reproduced by Waterlot[2] are by no means contemptible, but are well-balanced designs, interestingly treated. The bronze plaques of Benin, which covered the walls of the royal palace, have long been some of the best known and most popular forms of African art, chiefly because their air of sophistication and their representation of Portuguese types bring them nearer to our understanding. Whatever their merit, to judge all these explicitly as decorative panels is to use the wrong criteria; although some of them, through their planned arrangement and surface pattern, may be considered in this light. West African low relief carving in wood, used chiefly as door panels but sometimes also set into walls, approaches far more nearly to our definition of design. In these panels the stylized figures of men, beasts, and mythical beings, half-human and half-animal or -reptile, form definite though unsymmetrical

---

[1]Schmalenbach, *African Art*. Macmillan, New York, 1954, p. 154.
[2] Waterlot, *Les Bas-Reliefs des Bâtiments Royaux d'Abomey*. Paris, 1926.

patterns, often contained within bands of geometrical design. In some panels the pattern may consist entirely of interlacing bands or other geometrical motifs.

In both the bronze plaques and the wooden panels the use of small surface pattern gives a richness of texture to the work as a whole. Often the background is practically undecorated, in metal work it may be just relieved by lightly engraved trefoil or quatrefoil forms; in the case of wood the slight unevenness of the tooling prevents monotony; but the figures and other objects represented in relief are a mass of pattern, usually following round the form in bands.

Abstract Design in Moulded Relief
(*Plates VI, VII, VIII*)

Moulded pattern, as apart from the naturalistic representation of form, is often found covering the whole surface of external or interior walls and ceilings. In the huts of the Hima of Uganda the motifs used are symbolic, although in many cases their meaning has been forgotten today. These symbols are scattered in a somewhat haphazard fashion over the surface of the wall, but nevertheless produce a pleasing decorative effect with their contrasting colours of red ochre, black and white. On the mud walls of the houses of some parts of Northern Nigeria all types of moulded decoration may be found from pattern wandering across a whole external wall with no suggestion of a repeat, to sets of separate motifs firmly fixed within regular panels.

Painted and Moulded Ornament on the Surface of Walls and Ceilings
(*Plate VIII*)

The highly ornate vaulted ceiling at Kano illustrated here, is a most advanced example of moulded pattern, and it is interesting to notice the use of coloured china plates inserted as bosses at the junction of the arches. This use of plates or bowls, or even the necks and bottoms of gin bottles sunk deep into the walls so that they appear flush with the surface, is common in many parts of Africa where Moslem influence is strong; such as the Sudan and neighbouring territories to the South, and also the East Coast.

Further use of extraneous material is made by some tribes who construct mosaics of broken pottery and shells, and others who form geometrical patterns with strips of cane set into the wall.

DECORATIVE REED WORK
(Plate IX)

A less permanent type of material, but one which is nevertheless used in wall decoration by certain tribes is raffia palm, cane or reed. The Inter-lacustrine Bantu tribes of East Africa lined the walls of their huts and covered the interior supporting pillars of their roofs with reed-work, usually made of cane or elephant grass. These, bound with dark lines of bark, were very decorative, and the work in the houses of chiefs was of a high standard of craftsmanship. The highest development of this craft was to be found in the screens used by the Tusi of Ruanda-Urundi to partition off that part of the hut where the sacred milk pots were kept. These screens, patterned in black on the natural colour of the reeds, were decorated with a number of traditional symbolic motifs, very largely triangles and zigzags, fitted together in various ways.

The Bushongo, of the Belgian Congo, covered both the interior and exterior walls of certain huts with patterned matting. The patterns are geometrical and contain many of the motifs used in the famous pile embroideries of the tribe. It is not clear exactly how this work is done, as Torday and Joyce[1] speak of a frame-work of wood, to which the midribs of raffia palm are fixed horizontally. These are then bound with leaf-fronds, some dyed black, some in natural colour. Van den Bossche,[2] on the other hand, describes the weaving, without a loom, of separate mats, with a warp of mid-rib and weft of twisted cord made from the leaf-fronds; this is woven in twined weave. The finished mat is then fixed into position on the framework.

Thus colour, clay, wood, metal and fibre, all the materials known to a pre-industrial people, were turned to account in the beautifying of the walls of buildings in various parts of Africa. The satisfactory use of any material depends on the understanding which the craftsman has of its possibilities and limitations, for each material has its own peculiarities. That the African craftsman had such an appreciation of appropriate treatment is clear from a study of his varied methods of wall decoration, in which a wider variety of materials and techniques may be used than in any other craft.

---

[1] Torday and Joyce, 'Les Bushongo'. *Annales du Musée du Congo Belge*. Series III. Tome II. Fasc. I.
[2] Van den Bossche, 'Art Bakuba'. *Brousse* 1952. No. I.

CHAPTER III

Patterns on Mats and Screens

After our consideration of decorative reed and palm work in the construction of walls in African huts we come naturally to matting and screens made from the same materials, for matting fits halfway between these fixed woven partitions and woven textiles. Both the materials and the techniques used are essentially the same in all three crafts, the materials employed in mat making being finer and more pliant than those required for partitions, yet less fine and supple than those used for textiles. This similarity in construction results, as might be expected, in similarity of pattern. It is, in fact, difficult to draw a hard and fast line between one craft and another, and for the sake of convenience certain partition screens made by the Tusi of the Eastern part of the Congo have been illustrated in both this section and that on wall decoration, while the finely woven raffia mats or cloths of the Central and Lower Congo have been placed among the textiles. These Congo mats are woven on a well-developed type of loom, but more usually only the most simple form of frame is used for holding one set of fibres, and more often than not the work is done with no loom or frame at all.

Technically speaking the mats of negro Africa seem to fit into three large groups; there are those which are woven with or without a loom, others which must be described as sewn rather than woven, and a third class which are made from strips of plaited palm.

WOVEN MATS
(Plates X, XI)

The techniques shortly to be described for textile weaving hold good for woven

23

matting, with the proviso that the mat probably gets harder and rougher wear, so that the strands cannot be safely floated over a large number of others if they are not to be rapidly broken through use. Mats with the warp in one or two colours, and a weft of natural colour, with a large-scale geometrical repeat of a fairly complicated nature are typical of the work of the Lower Congo. Often the work is in plain black or white, at other times rather more colour may be introduced.[1]

Another type of mat from the same area, with animal, bird or human motifs can be most entertaining and attractive. These mats are woven with a warp of white crossed by a weft of black in such a manner that certain large areas appear predominantly white and others black. In this way the shapes of the animals and other motifs are clearly defined either as a black mass, or as a solid form decorated further with small pattern units. Of the three examples shown in Plate XI the first has been included, not because of any merit in the design, but because it shows clearly the technique employed. In both the other two a better sense of design and repetition can be felt, which is especially interesting in the bird pattern.

SEWN MATS
(Plates IX b, X b, XII)

The type I have termed sewn matting has no technical affinity with weaving, but is worked in a similar fashion to coiled basketry. A set of thick rigid splints is sewn together with raffia or some similar material, the fibre passing round the latest splint and piercing the edge of the one below, which is already in place; the stitches being fitted close together so that they completely cover the splint. Almost any type of pattern can be produced in this technique.

The more complicated mat and two screens shown in Plates IXb, and XIIb, also belong to the class of sewn matwork. In the mat the fine parallel splints of the ground are sewn in place by the bands of black pattern. Further interest is given to the design by means of variation in depth, for the pattern is stitched over a fibre cord and so raised like piping in European embroidery.

The screens are examples of the work of the Tusi, which is outstandingly beautiful in its simplicity of design. These people use a comparatively small number of motifs in their mat-making and basketry and restrict their colouring to black on a natural ground, with the occasional addition of a brown. The workmanship is always extremely fine and delicate, and this, together with the quality of the design, gives results of a

[1] See *Annales du Musée du Congo Belge*. Series III. Tome II. 'Céramiques et Nattes'.

very high standard. Technically these screens have a woven groundwork of flat splints supporting a patterned covering of black and natural coloured reeds, which are held in place by rows of thin fibre threads twined between each reed. The fastening forms no part of the pattern and is kept as invisible as possible.[1]

PLAITED MATS
(*Plates XIII, XIV*)

Finally we have plaited matting. Mats of this kind are made by joining together long strips of plaited palm fronds, dyed in various colours. The fronds are finely shredded and any number of strands from six to over fifty may be used in a plait at any one time. The pattern is produced both according to the order in which the different coloured strands are led in from the sides and by the numbering of the opposite strands over and under which they are passed. Each line of plaiting may be made to a different count after the fashion of a knitting pattern. The finished strips are finally sewn together to form a complete mat.[2]

It would seem probable that the craft was introduced into Africa by Arabs from the East, although it has now penetrated many hundreds of miles inland. On the East Coast and the offshore islands it may be seen at its best and gayest, while inland only the most simple patterns are usually found. In a collection of over sixty patterns recently obtained by the writer in Zanzibar the motifs are mostly named according to a fancied or obvious resemblance to a natural object. The Snake, the Bow tie, the Dog's ribs, the Fish, are typical titles. Others such as the Face of the Unmarried Woman, or the Place of the heart, are more exotic. The two prayer mats, also from Zanzibar, illustrated in Plate XIV, are extraordinarily beautiful specimens. The broad strips of pattern would entail working with some fifty strands in the hands, and the irregular pattern formed by the use of Arabic characters in the lower mat would require most skilful counting. The actual design of the upper mat has something of the quality of the design of a Persian rug.

Plaited mat patterns have also found their way into East Africa by a rather different route, being made by the wives of Nubian soldiers from the Sudan. A collection of patterns from Nubian matting is shown in the lower half of Plate XIII. It will be seen that these differ considerably from those shown above them, although in actual fact the difference is only due to one technical change. The Nubian women plait their mats

[1] See Czekanowski, *Forschungen im Nil-Congo- Zwischengebiet*. 1907–8, Vol. II.
[2] See Trowell, *African Arts and Crafts*. Longmans, 1937, pp. 121 et scq.

with a double thickness of strands, working with one colour over another, so that when the end of a plaited line is reached and the strands bent back to be worked in the opposite direction the colours are reversed. The interest of Nubian matting depends on this colour factor rather than on the use of complicated motifs, and one very simple pattern can be found in countless mats worked in various combinations of colour changes.

CHAPTER IV

Textile Design

WOVEN CLOTH
(Plates XV, XVI, XVII, XVIII)

Many early explorers of the West Coast of Africa state that they found the craft of weaving was highly developed there when they arrived. Ling Roth quotes a number of these in early accounts of Benin. First Welsh,[1] one of an English party who went to Benin about 1590, who speaks of 'cloth made of cotton wool very curiously woven and cloth made of barke of the palme trees', and 'The pretie fine mats that they make'.[2] Then an unknown Dutchman, D.R., in 1604 writes of 'much thread spun from cotton, of which they make their clothes similar to those on the Gold Coast, but lighter and finer'.[3] We get a further idea of the extent to which the craft was practised from Nyendal, who in 1704 says, 'not only all the inhabitants are clothed with it, but they annually export thousands of woven cloths to other places',[4] and Landolphe in 1778 writes 'few houses are to be seen without a cotton spinning machine, or frame for making admirable cotton or straw rug'.[5] Then Herskovits[6] quotes a legend which tells of the introduction of weaving into Dahomey in the second half of the seventeenth century by one of the kings, an accepted culture-hero of the tribe. In the Congo, too, the craft was evidently practised from early days, for Pigafetta,[7] writing of the Lower Congo in 1591, speaks of 'garments made from the palm tree of beautiful workmanship'.

[1] Ling Roth, *Great Benin*. Halifax, 1903, p. 131.   [2] Ibid. p. 140.   [3] Ibid. p. 140.   [4] Ibid. p. 141.
[5] Ibid. p. 141.
[6] Herskovits, *Dahomey*. New York, 1938, Vol. I, p. 16.
[7] Filippo Pigafetta. *History of the Kingdom of Congo*. A report of the Kingdom of Congo and surrounding

The weaving of textiles has been practised throughout West Africa, the Congo, and parts of Portuguese East Africa and Tanganyika. It is probable that the technique of weaving on a loom developed from that of mat making without such an apparatus after it had been realized that the work would be much simplified if ways were found, first of fixing all the warp threads at a certain tension so that they neither tangled nor worked loose as the weaving proceeded, and secondly of making sheds or openings through which the shuttle or needle carrying the weft thread could be passed in a more simple fashion than by picking the warp threads up separately each time with the fingers. Various forms of loom are used and these have been described in general by Ling Roth[1] and in more detail in monographs on various tribes. In order to understand the possible range and variety of pattern making on primitive looms such as are used in Africa, a certain amount of technical knowledge of weaving is required; but this is not the place to enter fully into details of the actual technique, and it must be taken for granted that the reader has some familiarity with the principles of the craft or can read it up in one of the many books on the subject.

Various fibres are used in African weaving. In the West Coast and the Congo finely split unspun raffia is very common, screw-pine is also said to be used in parts of Nigeria. Cotton is common on the West Coast and also silk, while a mixture of silk and cotton is used in some pattern weaving in Nigeria, and raffia and cotton in Dahomey. These variations in material add to the interest of the textures of the cloths produced.

Choice of colour is, or was before the introduction of synthetic dyes, dependent on the material used. Raffia was dyed with vegetable dye and most commonly patterned in black or red on a natural ground, but within the colour range could be found all shades of red and yellow ochres, siennas and sepias, greys, lavender-pinks and faded yellows. Cotton in Nigeria is dyed with local indigo, giving a range of blues, but imported dyes are also used. Ghana silk weaving is often a mass of brilliant colour, and it is said that the first imported silk cloth from Europe was unpicked and rewoven by Ashanti weavers in order to incorporate the bright new colours in their own designs.

The types of pattern which can be made in weaving can roughly be divided into two groups. First come those which are produced in a piece of plain weaving simply by the arrangement of different coloured threads in the warp or weft, or both. Coloured threads in the warp will result in longitudinal stripes of which the number, width and colour are arbitrary; while, when they are used in the weft, transverse stripes will be produced across the width of the material. If both transverse and longitudinal stripes

countries; drawn out of the writings of the Portuguese Duante Lopez, in Rome. 1591. Newly translated from the Italian and edited by Margareta Hutchinson. London, 1891.

[1] Ling Roth, *Studies in Primitive Looms*. 3rd edition. Halifax, 1950.

are introduced the resulting pattern may be in the form of a check. A great variety of striped and checked pattern may be obtained in this way. If, for instance, red, green and white are used in both warp and weft, combinations of red crossed with red, or green, or white, and green crossed with green or white, and white crossed with white will give six variations in colour. If there are more warp than weft threads to an inch the weft will be largely hidden and the weave is called warp faced; if the position is reversed and the warp hidden by the greater number of weft threads it is known as tapestry weave. Then the width of the different bands of colour may be varied, and a further interest added by the use of threads of varying thicknesses, or even of differing fibres. Plate XVa gives some indication of such simple pattern.

More complicated forms of pattern weaving are introduced by allowing certain threads to pass over one or more than one thread in either warp or weft. To do this the warp threads are sometimes set up in pairs together, and only the ones needed for the pattern are used, the others are floated at the back until needed. Alternatively a double weft is used, and the coloured one forming the pattern is floated on the right side of the material directly above the plain coloured one which is woven in to form the ground. It will be woven right across the cloth, or worked backwards and forwards over certain small areas as the pattern requires. The pattern is then practically invisible on the wrong side. In the Congo patterned weaving is often carried out in natural coloured raffia only, forming what is known as diaper, or plain coloured pattern.

By these technical means an unlimited number of patterns can be woven, and in-finite variations on simple geometrical designs are found in African cloths, although superficially they conform remarkably to type. The motifs are most often geometrical, but will range from broad stripes to delicate diagonal and zigzag lines; other favourite forms are solid triangles and lozenge shapes; while in some cloths more or less stylized human or animal forms can be found. A very characteristic arrangement in African textile design, on either woven or printed cloth, is the division of the material into bands or panels which are each filled with repetitions of one particular motif. These bands or panels are sometimes separated from each other by borders of stripes or plain uncoloured weaving.

EMBROIDERED CLOTH
(Plates XIX, XX, XXI, XXII, XXIII)

Both barkcloth and woven textiles may be further decorated with various forms of embroidery. It is not always easy to gather from the writings of the early explorers just what was found in Nigeria when they first arrived, although several accounts would suggest embroidered as well as woven material. It would be interesting to see, for instance, the hangings in the royal buildings at Benin. Fawckner (1825) tells of 'a large hall hung with the most superb cloths of the country like tapestry'. Burton (1862) talks of seeing in the same place 'Fine cotton work, open and decorated with red worsted'. And Punch (1890) describes 'pieces perhaps six feet long or more, with life sized figures worked with a needle on the open-worked cotton material'.[1]

The most famous, and certainly the most beautiful, African embroidery comes from the area around the confluence of the Sankuru and Kasai rivers in the Belgian Congo, where it was carried out by the various sub-tribes of the Bushongo or Bakuba. The work is of such a high quality both in design and in technical achievement that it deserves special consideration here. It is said by Torday and Joyce[2] that the craft cannot be traced further back than the early seventeenth century among the Bushongo, and that the Mbala tradition was that Shamba Bulongongo, the most famous king of the Bushongo, learnt the art from the Pende and brought it back to his own country. Today it is considered that the origin of the craft may well have been the Lower Congo.

The embroidery was worked on a loosely woven, coarse-stranded raffia cloth which was first pounded in water or worked with the hands until it was as soft and supple as possible. This canvas was sometimes left the natural colour or sometimes dyed red with *Takula* (*Pterocarpus*) or blue with indigo. All shades of blue-purple, Indian red, or lavender pink can be found. Both ground and embroidery together were sometimes left a natural colour, or both dyed red after the work was completed.

The embroidery fibres were dyed red, yellow, mauve or black, or various shades of brown, or whitened with some mineral substance, or left the natural colour. With only simple vegetable dyes available the colour range was limited, but there were many fine gradations and the effect was subtle and beautiful. The raffia for the embroidery strands was split as finely as possible and worked in the hands until it was very soft; several threads were then twisted together and threaded through an iron needle.

The best known of these embroideries are the pile cloths (Plate XIX). In this type of work the shapes were filled in by threading the needle under one strand of the weave and pulling the fibre through until only about 2mm. were left on the far side of the stitch. The fibre was then cut off at the same height on the near side, and when this stitching had been repeated closely over the whole area, the result was a smooth,

---

[1] Quoted. Ling Roth, *Great Benin*, p. 141.

[2] Torday and Joyce, 'Notes ethnographiques sur les peuples communément appéles Bakuba'. *Annales du Musée du Congo Belge*. Series III. Tome II. Fasc. I. Brussels, 1910, pp. 183 et seq.

Further reference. Torday, *On the trail of the Bushongo*. London, 1925.

unknotted, brushlike surface resembling a pile velvet. The work was so finely done that the embroidery fibres did not appear on the wrong side. Various different ways of outlining or filling the motifs were employed, giving interesting changes of texture. Sometimes the forms were outlined in black stem stitch, sometimes several rows of stem stitching in different colours surrounded the central filling. Sometimes the filling was not continuous, the pile being worked as separate dots without continuity, sometimes it was cut very short, sometimes longer, and sometimes not cut at all, being left in loops.

To the west of the Bushongo area, particularly among the Mbala, more ordinary forms of embroidery were worked as well as the pile cloths (Plates XX, XXI). The stitch used was either stem stitch or a complicated form of chain stitch, while in some of the early eighteenth century Mbala cloths a type of drawn thread work was also done. The raffia canvas used for this work was of a very fine quality and the embroidery most delicate. Sometimes the ground was dyed black and the designs embroidered in natural tones upon it, sometimes the whole work was dyed red when finished. The pattern motifs used by the Bushongo are discussed in a separate section. Embroidery was always a woman's craft.

Other embroideries to which reference must be made are those of the Hausa, Fulani, and other peoples of Northern Nigeria. Here robes and curiously shaped wide trousers are covered with the most elaborate pattern work in coloured and gold and silver thread. Plate XXIIa shows a part of a beautiful scarlet robe embroidered with silver and gold from Bida in Northern Nigeria; the detail is very finely done, and the total effect of the robe, with the front panels and the back of the shoulders embroidered, is extraordinarily rich. The embroidered robes in Plate XXIII are of a rather different type, being far less elaborately covered than those illustrated above. The decoration of the portion of the wide trousers illustrated in b consists of a number of motifs put together in a somewhat haphazard way, very different from the regular balanced work of the previous examples.

PATCHWORK

Patchwork is not a common method of decoration in Africa, and the few pieces which can be found in the museums are old cloths worn by the Bushongo women at certain ceremonies.[1] They are made from a number of small triangular, rhomboid, or rectangular pieces of barkcloth, some dark brown or black, others a light natural

[1] Torday and Joyce, ibid. p. 180.

colour sewn together with fibre. There is nothing very striking about them, but it is interesting to note that this method of pattern making with textiles was indigenous in Africa.

APPLIQUÉ
(Plates XXIV, XXV)

Appliqué, or the ornamentation of cloth by stitching on either pieces of fabric of another colour, or completely different material such as beads, cowrie shells, metal plates, and so on, is found among the West Coast tribes; and also, albeit in a very different style, among the Bushongo of the Congo. In the great kingdoms of the West Coast its use seems to have been restricted to ceremonial functions or associated with civil or military power. Talbot[1] talks of 'the old war banners of the Kalabari tribes' being of appliqué work, while many of the symbols associated with the kingship and queen-mothership of the Ashanti collected by Meyerowitz[2] are from appliqué cloths, but most information about the use and technique of such cloths is given by Herskovits[3] in his account of the Dahomey.

These cloths were made by members of an especially restricted family guild for the use of those of rank and power. The craftsmen lived in a compound near the royal palace, and no such work was done anywhere in the country save by the men of this group only. We have accounts of the work from comparatively early days of British exploration, for Forbes describes it in 1849 and Skertchly in 1874.

Appliqué work was used for such things as state umbrellas, chiefs' caps, banners of associations, distinguishing marks of companies in the army and other groups; and for pavilions, hammocks, and hangings in chiefs' houses. The motifs were heraldic devices or symbols, or the insignia of kings; and the achievements of great men or incidents from everyday life were recorded in pictorial form. Herskovits believes that the designs used were derived from the bas-reliefs which decorated the walls of the royal palaces, and while there may have been no formal copying of actual reliefs the similarity of subject matter and the necessity in both techniques for large simple treatment of forms would lead to certain resemblances. The greater number of motifs were those of animals which were the symbolic forms used to set forth the greatness of the king and the royal dynasty.

[1] Talbot, *Tribes of the Niger Delta*. London, 1932, pp. 276, 335.
[2] Meyerowitz, *The Sacred State of the Akan*. London, 1951. Chap. VI.
[3] Herskovits, *Dahomey*. Vol. II, pp. 329 et seq.

According to Meyerowitz[1] the craft was first learnt in Dahomey from a Brazilian Portuguese who had settled in Widah. It is said that the work was originally done on raffia cloth, but that European trade cloth had been used since 1890. The background was usually a brilliant gold or black, figures were done in red or black; while purple, blues or greens were the predominating auxiliary colours. Patterns of individual figures, traditionally stylized, were kept for regular use, but motifs and forms altered from generation to generation of workers in the craft. Technically appliqué necessitates the main motifs in the design being worked in bold silhouette, with details such as features and the texture of hair, scales or feathers, or the patterning of clothing, indicated in a conventional manner by stitchery.

Painted Cloth
(Plates XXVI, XXVII)

The weaving of patterns, or the decoration of plain cloth by covering it with embroidery, or by sewing on pieces of other material of differing colour and texture does not by any means exhaust the possibilities of textile design, for another vast field is opened up through the use of paint and dye.

Obviously the first form of painted decoration on cloth was made by the straightforward method of applying the paint with a finger or some simple kind of brush; making a pictograph to tell a story, or marking the cloth with a symbol or identification sign, or ornamenting it with a pattern. Decorated cloths of this kind can be found in many parts of Africa, the work being done on both woven fabric and barkcloth. It is often crude and clumsy and lacking in any real sense of design, but some cloths from the Congo painted in this way in greys, blacks and browns on a light ground are very beautiful. Indeed the almost complete freedom from any technical considerations in this kind of work makes for most spontaneous and lively results.

In India a highly skilled technique of outlining a design on the fabric with a reed pen and painting in the colours by hand was developed. It is interesting to note that a very similar method was used in Ghana. In the British Museum are two specimens of painted decoration on imported white cotton drill. The outlines of the pattern appear to have been drawn with a pen in brownish-black and certain areas are closely filled with scroll-like linework resembling Arabic characters in the same colour. Other smaller areas are solidly painted in green, red or yellow. The work has none of the finish or delicacy of Indian textile design, but it is even more remote from the simple broad

[1] Communication.

33

irregular treatment of other specimens of African painted pattern illustrated here. It has much in common with the fine embroidery on the robes of Moslems in the northern territories of the West Coast, and would appear to be an attempt at imitating this work in a cheap and rapid fashion.

PRINTED CLOTH
(Plate XXVIII)

In the decoration of textiles the desire for the regular repetition of the pattern unit over a large area calls for some mechanical method of reproduction which will be quicker and more precise than free painting. Some tribes cut stencil plates from the thick green sheaths of the stem of the plantain, or print with simple units cut on a cross section of a stick. But the finest examples of African skill in printing on cloth are the *Adinkira* stamp-printed cloths made by the Ashanti. The stamps used are cut from small pieces of calabash, and a very large number of different motifs are used. Rattray[1] gives drawings of some fifty of them, with the names by which they are known. As even in a large calabash the curve of the skin is perceptible it is not possible to cut a piece sufficiently flat for a stamp larger than three inches in diameter, while most stamps are considerably smaller. Several long slivers of bamboo are stuck into the back of the stamp and tied together a few inches from the base, forming a handle. The motifs are sometimes cut right through the thin calabash skin, or sometimes left in low relief. The cloth used is of woven cotton, sometimes left white, sometimes dyed russet brown with the bark of a tree. The black dye used for printing is prepared from another kind of bark which is cut up and boiled for several hours together with lumps of iron slag. After two thirds of the water has evaporated the remainder is strained off and the dye which is left is of the colour and consistency of coal tar. The cloth is pegged out with small wooden pins on a piece of flat ground for printing and the stamp then dipped in the dye and pressed on to the cloth.

In the type of stamp printing described above, the stamps have been designed as separate units each complete in itself, and are usually so used; they do not normally form part of a continuous flowing pattern as do blocks used for printing in both Europe and the East. Arab influence along the East Coast of Africa brought this different approach to the craft and at Zanzibar wood blocks were being used by the Swahili for printing *kangas* in the early part of this century which could be repeated to form a continuous pattern with no signs of a join. Some of these blocks were cut in

[1] Rattray, *Religion and Art in Ashanti*. Oxford, 1927. Chap. XXV.

34

low relief, others had groups of wooden pins stuck into the face of the block by which spots were printed. Many of the actual motifs used, as well as this difference in technique, betray the Eastern origin of the craft in this part of Africa.

Dyed Cloth, Tie Dyeing
(Plate XXIX)

Pattern printing with a stick, stamp, or block produces a darker motif on a light ground, other processes achieve the opposite effect by so treating the pattern areas that, when the whole cloth is immersed in the dye bath, they are left light while the rest of the fabric takes up the colour. There are two main methods of doing this; one, known as tie dyeing, consists in tying or sewing certain parts of the cloth so tightly together that the dye is unable to penetrate them; the other in covering the pattern units with either a wax or a paste, so that the dye cannot sink in. This is called the Resist method or Batik. Both crafts are practised in Africa, chiefly in Nigeria.[1]

An examination of specimens of African tie dyeing shows several variations in technique which produce different effects. A narrowly woven strip of cloth may be gathered horizontally at intervals and the gathering fibre drawn tight, after which raffia is bound tightly round the gathered portions so that the prepared material resembles a rope. When opened out after dyeing the cloth shows a series of horizontal undyed lines, varying according to the closeness of the gathering and the thickness of the material. When the cloth is caught up into little peaks bound tightly with raffia the result after dyeing is a white spot on the dyed cloth. A skilled craftsman can so twist his protecting fibre that the spot becomes a square or an oblong. If the cloth is bound round with parallel bands of fibre below the peak a number of concentric rings of light will result. Seeds or small stones can also be tied into the cloth to produce a spotted effect. Although the shapes which can be produced by this process are rather restricted, consisting simply of dots, circles, and parallel or zigzag lines running in one direction only; or, through the gathering of many little bunches of stuff close together, a splashed effect; yet very attractive cloths can be produced by the grouping of the pattern forms.

A modification of the tying technique is obtained by stitching the cloth. A pattern is sometimes carefully embroidered on a cloth with finely stranded raffia, and after dyeing the stitching is completely unpicked leaving the light ground showing. A chequered texture can be produced by darning the cloth with fibre, and animal or other forms are sometimes picked out in this way. The Jukun of Nigeria outline a design

[1] 'Art in the Drying Field'. *Nigeria* 30. 1949, pp. 325 et seq.

on the cloth by stitching strips of palm fibre along it before dyeing,[1] while the Bushongo of the Belgian Congo sew small pieces of reed or cane over the parts of the cloth which they wish to protect from the dye.[2] The Yoruba and the Tiv of Nigeria make many tie-dyed cloths, using the local indigo for their dye.[3, 4]

The Musée de l'Homme possesses one or two very interesting pieces of tie dyeing from the Ivory Coast. The material used is very finely woven raffia cloth which would appear to have been tied and dyed twice, the colours being the natural light yellow ochre, a darker red ochre, and black. The interest of the specimens lies in the fact that after dyeing the cloth has been deliberately left puckered and crimped instead of being stretched out and flattened, so that it has a curious spongy elastic texture.

Nigerian craftsmen are aware of the beauty which small irregularities of texture or colour can give to a work, for it is said of the tie dyeing method that sometimes red cotton is used for the sewing; this stains the cloth and, although it will later wash out, gives a finish which may help to sell it, just as does the indigo which is often applied superficially to cloth at the end of the dyeing process in order to give a temporary sheen which is valued by the buyers.

DYED CLOTH, RESIST METHOD
(*Plate XXX*)

The methods of sewing reeds on to the cloth or embroidering it with cotton or other fibre threads leads on to resist printing. In true resist printing the parts of the cloth which are not required to take the dye are covered either with wax or with a starch paste which will resist the penetration of the dye stuff into the material.

Amongst the Yoruba two different methods are practised. In one starch paste is painted on to the material directly with a feather or other tool. In the second a zinc-foil stencil is cut, clearing away the parts which are to be left white on the cloth. The stencil is then placed over the fabric and the paste painted through it with a piece of wood. Tin or even leather is also used for a stencil plate. After the paste has been put on to the cloth by either method it is dried until hard. The cloth is then dyed and after drying the surface paste is flaked off and the remainder finally boiled out.

[1] Meek, *Northern Nigeria*. Vol. I, p. 163.

[2] Torday and Joyce, 'Notes ethnographiques sur les peuples communément appelés Bakuba, ainsi que sur les peuplades apparentés les Bushongo'. *Annales du Musée du Congo Belge*. Series III. Tome II. Fasc. I. Brussels, 1910.

[3] Mrs. Daniels, 'Yoruba Pattern Dyeing'. *Nigeria* 14. 1938, pp. 125 et seq.

[4] K. C. Murray, 'Tiv Pattern Dyeing'. *Nigeria* 32. 1949, pp. 41 et seq.

It is said that there is no evidence that stencilled resist pattern making among the Yoruba has been introduced by Europeans, but it is a craft which is still developing and the motifs used are not necessarily traditional but are sometimes representations of men, animals and objects. Older pieces of resist printing which have been collected are usually strictly geometrical in their patterning, although some have simple stylized representational forms. Imported cotton goods from Manchester striped in bright reds, yellows and other colours, and also floral chintzes, were often over-printed locally in indigo by both the resist and tie dyeing methods so that in the lighter portions protected from the final dyeing, the original colours and patterns were left exposed. This often gave a very interesting effect to the cloths.

By the use of the stencil the treatment must in some ways be more limited than in freely painted work, because as in all stencilling, the solid portions of the plate must remain tied together, thus breaking the pattern units. One specimen in the British Museum has a fish scale pattern with a number of concentric curves on each scale: the cloth may have been covered with paste and the pattern then combed off.[1]

DYED CLOTH, THE DISCHARGE METHOD
(Plates XXXI, XXXII)

In this method of pattern printing the whole material must first be dyed. The design is then produced by printing or painting on the prepared background with a powerful reducing agent which removes the colour and so leaves a white or pale coloured pattern on a darker ground. The technique was not used in Europe until the beginning of the nineteenth century, and would seem to require considerable scientific or empirical knowledge.

A collection of cotton cloths printed by the people of French West Africa and particularly the Bambara was brought to the Musée de l'Homme by the late Fr. de Zeltner in the early part of this century. In his description[2] of the technique used he clearly describes a discharge method, resulting in white designs on a fairly dark ground. The woman who dyed the material first prepared a dye bath by boiling the bark or leaves of certain trees for three hours. The cloth was immersed in the brownish liquid for about a day and then rinsed. Then, using a piece of iron as a tool, the design was painted on in a special mud, probably containing iron acetate, which was found at

---

[1] See Mrs. Daniels, 'Yoruba Pattern Dyeing'. *Nigeria* 14, pp. 125 et seq.; 'Art in the Drying Field'. *Nigeria* 30, pp. 325 et seq; Wenger and Beier, 'Adire, Yoruba pattern dyeing'. *Nigeria* 54. 1957.
[2] Clouzot, *Tissus Nègres*. Paris.

the bottom of certain pools. When dry the pattern was gone over again with locally made soap on the iron tool. This soap was obtained by soaking ashes in certain vegetable oils and it contained a good deal of potash which acted as a mordant. The pattern was then covered once more with mud and dried in the sun. Finally the cloth was beaten with a rod, rubbed between the hands, and rinsed in water to remove all traces of mud, and the pattern stood out white on the dark ground.

The patterns which can be produced by both the resist and discharge methods are somewhat similar, as both allow for free drawing on the cloth. As may be seen from the illustrations, very fine clear-cut detail has been obtained in the Bambara discharged cloths, although from the description of the covering of the patterned portions first with mud, then soap, and then mud again, it would seem extraordinarily difficult to avoid thick, blurred lines.

A study of a number of these cloths from slightly different areas shows interesting differences and similarities in motifs. Some rely entirely on the effect of linear patterns, others use more solid areas set against portions broken with fine line. Certain motifs can be found repeated on more than one cloth. The general effect of the cloths is of a more disciplined type of pattern than that found on the resist printed cloths of Nigeria; and considering the great technical skill required to get clear-cut lines in work of this type they reach a very high standard indeed.

CHAPTER V

Ornamental Basketry

It would be impossible fully to cover the infinite variety of pattern which can be found in African, or any other, basketwork. The very techniques of the craft themselves with all their variations produce a large number of patterns in texture over the surface of the baskets, and these may be further exploited through the development of the possibilities of each type of weave by the introduction of variations in materials and colour.

WOVEN BASKETRY
(Plates XXXIII, XXXIV, XXXV, XXXVI, XXXVII)

The type of pattern motif which is possible in woven basketry is similar to that found in the related techniques of cloth weaving and matting. Although in a large number of cases the pattern is woven in the body of the basket itself, it is obviously impossible to produce delicate and detailed patternwork with the tough and comparatively thick fibres and splints which are necessary for strength of construction. For this reason the finest examples of woven pattern are found almost without exception in small baskets which are first constructed of bark or of woven fibres and reeds of the required toughness and then covered with a very finely woven decorative outer case. In this way most beautiful specimens are made both in the lower Congo and in parts of Rhodesia.

In some techniques baskets are plaited rather than woven in a similar manner to the plaiting of strips for mat making.

ORNAMENTAL BASKETRY

COILED BASKETRY
(*Plates XXXIII, XXXV, XXXVI*)

In some parts of Africa coiled basketry is more common than woven. In this technique a central core of grass or reed is sewn with fibre thread in a flat or ascending spiral. The core itself is usually invisible, though not always so, and the pattern is produced through the colouring of the stitching. Often the colour scheme is restricted to black or brown or red with the natural coloured fibre; or, as in the case of baskets and trays made by the Nubian and Swahili women of East Africa, the work is a mass of brilliant colour produced by imported synthetic dyes. In this form of work equally interesting pattern can be produced on a large as on a small scale.

BASKETS SHOWING VARIATIONS IN MATERIALS OR TEXTURE
(*Plates XXXVI, XXXVII*)

Finally a random search through the basketry from Africa in any large museum collection will reveal a number of baskets in which the decorative interest lies either in the combination of several different techniques, or in the use of two or more very different types of fibre, or the combination of woven basketry with some entirely different material such as wood. This emphasizes the African's great love of texture as apart from colour or form, which becomes obvious in the study of his craftsmanship in any material.

CHAPTER VI

Beadwork

(Plates XXXVIII, XXXIX)

Apart from the use of beads in the making of head, neck and facial ornaments they are largely used in West, East and South Africa for the making of patterned covers for all kinds of things such as pots, calabashes of all sizes, drums, staves, fetish objects, masks and the like. They are also sewn on to cloth to decorate belts, boots, bags, robes, and other articles of apparel.

In both West and East Africa beadwork is especially associated with the decoration of chiefs' robes and crowns; or with the covering of objects connected with royal regalia or of other things held sacred in the same way. As would be expected, much of the design has a symbolic significance, and complete appreciation of it would depend upon knowledge and understanding of local tradition and symbolism. Superficially the technique lends itself best to the making of geometrical motifs such as triangles, zigzags, and interlacing pattern. Trade beads are usually employed, the resulting colours being bright and strong. These are sometimes arranged with real appreciation; but often, especially in modern work, are no more than a dull combination of dark blue, red and white.

CHAPTER VII

The Decoration of Hides and Leather

Skins, both in the form of raw hide among primitive peoples, and dressed leather among the more technically advanced, have always been very widely employed. Not only can they be used for clothing, but also to make screens and shelters, harness and binding thongs, shields and scabbards, shoes, sandals, bags, water pots, and containers of all shapes and sizes. It is therefore not surprising to find that a number of different methods have been evolved for their decoration.

SHAVEN PATTERN
(*Plate XL*)

In East Africa some tribes decorate skins from which the hair has not been removed, by shaving it off in certain parts; they may either clear small solid areas in this way, or form the pattern with narrow shaven lines.

PATTERN CARVED IN LOW RELIEF
(*Plate XL*)

Thick, tough hide can be carved in low relief in very much the same way as wood or calabash, and this treatment is sometimes used to ornament sheaths and scabbards. If a dark hide is used, and the cleared portions of the pattern are rubbed with a white chalky material, they will stand out clearly.

42

THE DECORATION OF HIDES AND LEATHER

In fine leather work parts of the surface leather may be excised to form small checkered designs.

APPLIQUÉ WORK ON RAW HIDE
(Plate XL)

A yet more ornate method of decorating raw hide was that carried out on the fans used in Benin. To quote C. Punch[1] (1889) 'It was always "quite the thing" for an African potentate to have at least two slaves fanning him. The usual shape of Bini fans was . . . (as illustrated). They were made of cowhide (green) with the hair on, and bright coloured pieces of flannel or Hausa leather were sewn on to make the pattern. No Ukoba, starting out to make himself a nuisance to some village, would feel quite dressed without one of these fans—in fact, except for the anklet and necklet, the fan was all he took in the way of clothes.' Elsewhere we have reports of these fans, the dark hairy skin decorated with scarlet flannel appliqué, sewn in place with strips of thin yellow skin. Various motifs, including stylized human and animal forms, seem to have been used in their making.

PAINTED HIDE SHIELDS
(Plate XLI)

Perhaps the most exciting patterns on hides are the painted designs on shields, used by certain tribes in East Africa. The shields are painted in earth colours, usually red, black and white, and are most striking. The patterning of the two halves of the shield is more often than not asymmetrical, but is always well balanced and arranged with due regard to its basic shape. Amongst the Masai and kindred tribes a form of heraldry has been developed by which those who know the signs can tell the district from which the warrior has come, and whether he is considered by his fellows to be a brave man, and so on.[2]

---

[1] Quoted. Ling Roth, *Great Benin*, p. 123.
[2] See Hobley, *The Ethnology of the Akamba*. Cambridge, 1910, p. 127.

43

THE DECORATION OF HIDES AND LEATHER
DECORATION OF TANNED LEATHER
(Plates XLII, XLIII)

Turning to decorative work in tanned leather we find it to be the main traditional export of the peoples of Northern Nigeria as well as being extensively used in their own daily life.[1] In a description by Dodwell[2] of the leather work at Oyo in Nigeria it is stated that the skins used are almost exclusively goat skins; in the past sheep or various game skins were also used but never those of horses or cattle. Although imported dyes are now sometimes used the traditional ones are locally made. These consist of black from a mixture of pieces of old iron, starch, and ashes from the blacksmiths' workshops; red from the outer strippings of the bulrush millet together with salt and wood ash; yellow from a paste of ginger and lime juice; blue from indigo. For green and white the leather had to be untanned, when green was made from a mixture of copper filings, lime and copper sulphate; and white by rubbing in palm oil. A very similar list is given from Kano by Hambly.[3]

The most simple form of decoration in Northern Nigeria is said to be done by drawing arabesques on the leather with black ink, but patchwork and appliqué in which motifs in contrasting colours are first gummed and then sewn to the basic skin with narrow thongs are better known. The applied patches are further decorated with coloured stitching. In some cases checkered designs are made by the excision of parts of the surface leather.

The design motifs are traditional, and similar to those used in local embroideries with very few variations. As in most Moslem art they are largely geometrical, but stylized men, animals, birds and reptiles are also to be found as can be seen from the illustration of the cover from Bida (Plate XLIII).

Very similar work in leather can be found further south; it does not always have the same standard of technical perfection as work from the north, but may compensate by more original and interesting design.

[1] Jeffries, 'Leatherwork in Northern Nigeria'. *Nigeria* 14. 1938.
[2] Dodwell, 'The Tim-tim makers of Oyo'. *Nigeria* 42. 1953.
[3] Hambly, *Culture areas of Nigeria*. Chicago, 1935.

CHAPTER VIII

Cicatrization and Body Painting

(Plates XLIV, XLV)

A study of African pattern would not be complete without mention of the practice of cicatrization, and also of the painting of designs on the face and body with the juice of certain plants or with ochre or chalk. Such customs are dying out, but can still be found amongst the more unsophisticated people.

It is, after all, common to the human race and to women especially, to attempt to enhance natural beauty by some form of decoration. Amongst many people this is confined to costume or hair styles, or to painting such prominent parts as the face or finger nails, but the African carries the art considerably further. Living in a hot climate he has more exposed parts of his body to decorate, and he does this with the thoroughness with which he would decorate a calabash or any other vessel.

The African prizes an oiled and polished body, although sometimes he produces a matt surface by smearing it with clay or colouring matter. Both oiling and covering the body with colour are usually done for the dance. The hair is often trimmed and plaited into various styles, and certain teeth may be extracted or filed to a point.

In Nigeria some tribes ornament their faces and bodies with the most delicate and intricate patterns painted with vegetable juice, but the most usual form of decoration is cicatrization. There are several reasons for the practice; it is a common custom among a very large number of tribes to mark their members with the tribal mark, very often on the temples; then medicine, both curative and preventive, also employs this device, when the body may be cut and rubbed with medicine supposed to contain magical properties; but the most highly developed schemes of decoration are made for aesthetic

45

purposes, to give visual and tactile pleasure to young people in their more intimate personal relationships. Not every tribe which practises cicatrization carries the art to a high standard, often the work consists merely of a few rows of parallel scars, but, as may be seen from the illustrations, many highly elaborate and beautiful patterns may be found, especially among the tribes in the centre of the Congo. Here the most ornate designs are to be found covering the whole body.

'Il faut souffrir pour être belle', and the descriptions which are given of all forms of cicatrization seem painful and unpleasant. Both abstract designs and stylized forms are cut on all parts of the body—forehead, temples, cheeks, neck, legs and arms, chest, stomach, buttocks, thighs and legs; the marks are usually made by cutting the skin with a sharp knife or razor, and then rubbing charcoal or vegetable matter into the wounds so that they finally heal as raised, shiny keloids.

Bohannan[1] writes that there are changing fashions, both in motif and techniques, in Tiv cicatrization; and the decorations are used to emphasize the natural good points of the wearer's face or body. By using a technique which will make deep scarifications, for instance, prominent cheeks may be made more prominent; and the apparent shape of a nose can be altered at will. A good pair of legs, too, will attract more notice if tastefully decorated with a large and outstanding band of pattern. The Tiv take a great deal of trouble in trying out various patterns to see which fit their faces best. The young ones do this by painting with the juice of a certain plant, while a more lasting trial run may be taken by pricking the face with the twig of a tree which leaves a round white scar for two or three months. The actual process of cicatrization, with the list of patterns used in Dahomey with their names and sites on the body, has been given by Herskovits.[2]

[1] Bohannan, 'Beauty and Scarification among the Tiv'. *Man*. Vol. LVI, 1956, 129.
[2] Herskovits, *Dahomey*. Vol. I, pp. 291 et seq.

CHAPTER IX

Calabash Patterns

The decoration of calabashes is one of the most interesting fields for the study of African pattern. Amongst the tribes which have a strong penchant for ornament the craft gives great opportunities, and a number of different techniques have been evolved, each giving its own interesting effects. From the written accounts available we can get a clear idea of some at least of these processes, and it is usually easy to pick out the various methods used from a systematic study of the collections of African calabashes in the museums. From the subject matter which is found in the designs we see clearly the pre-eminent position which allegory and symbolism hold in African design.

The preparation of a decorated gourd or calabash is a long and painstaking process. First the gourd must be prepared. For this it is taken when ripe and soaked in a stream or water hole until the contents are completely rotten. It is then opened, and the insides thoroughly cleaned out, after which it is dried in the sun. During the soaking the skin will have grown bloated and soft, but as it dries it becomes hard so that it can be worked like wood. By some tribes the calabash is cut into two horizontally, the top portion often being used as a lid for the base; while others cut it across vertically. The different shapes thus produced lead to very different forms of ornamentation.

The next consideration is the basic colour of the undecorated calabash, and here considerable variety may be produced. The natural colour is a warm yellow, and it is often left in this state. If the fine layer of the cuticle or outer skin is scraped off the craftsman has a white surface to decorate. Alternatively the calabash may be rubbed with a concoction of millet leaves which gives it a beautiful old rose colour, or it may be stained indigo, red, or deep orange or burnt sienna. After decoration these various

47

colours receive a further rich patina through constant handling and through hanging in a smoky hut until they show a sombre polish. So even with no decoration at all the basic colours and textures of calabashes cover a large range and some of them are very beautiful in themselves.

CARVED PATTERN
(Plate XLVI)

The techniques which are then used in calabash decoration fall into four main groups, although in actual practice the different methods are often combined. First comes a broad treatment in which either the background or the pattern motifs are cleared evenly to a depth slightly below the surface giving clear relief and colour change. If the calabash is intended for purely ornamental purposes the whole skin, which is not very thick, may even be cut away in places.

SCRAPED PATTERN
(Plates XLVII, XLVIII, XLIX, L)

Closely akin to this carving method comes that of scraping. Many variations of decoration are produced by outlining the pattern motifs with a sharp knife and then carefully scraping away the background area. Thus if the calabash is unstained the pattern will stand out yellow (the natural colour of the surface) on a white (scraped) ground. If the calabash has first been stained the design will be the colour of the stain used, on a ground which is either yellow (the stain alone having been rubbed off and the natural colour of the cuticle left unharmed) or white (if the cuticle is also removed). A reverse effect can be obtained, of course, by scraping away not the background, but the pattern motif itself.

SCORCHED PATTERN
(Plates LI, LIId)

A third method of decoration is done by lightly scorching the surface area of the pattern motifs with the flat surface of a knife or other tool, thus colouring them dark brown or black. When whole areas are thus treated another colour possibility is added.

CALABASH PATTERNS

ENGRAVED PATTERN
(*Plates XLIX, L, LI, LII*)

Finally come the engraving techniques. In these, large surfaces, either of the background or of pattern motifs, are covered with a texture of finely engraved line, very often cross hatched. These lines are sometimes incised with a sharp knife, sometimes burnt with a hot point. When the latter tool is used a texture of small burnt black dots instead of lines may be produced. The engraved pattern will show up white or black according to whether it has been cut or burnt against the natural coloured or stained parts of the calabash, and this contrast is sometimes accentuated by rubbing soot or chalk into the engraved lines. The technique may be used in such a way that the whole background or the whole surface of the pattern motifs are covered with a mass of texture pattern.

DECORATION WITH EXTRANEOUS MATERIALS
(*Plate LIII*)

The ornamentation of calabashes by the application of extraneous materials is known. In both South-east Africa and West Africa soft clay is pressed into incisions and coloured or white beads embedded in this base, either as ornamental motifs in themselves or used to outline bands of scorched pattern. Gourds are commonly completely encased in tightly fitting coverings of patterned beadwork, constructed round them but kept entirely separate. Patterns are sometimes 'embroidered' on calabashes, closely stitched with brass or steel wire or even, from the Ashanti in Ghana, with gold thread.

All these methods of carving, scraping, scorching or engraving pattern on calabashes are most suitable both for geometric design and for the representation in stylized form of animals, men, and other objects. When geometrical motifs alone are used, rich contrasts in texture may be obtained by leaving some forms in strong colour contrasted with others in white or buff. Some calabashes are further enhanced by the engraving of small texture patterns on certain areas. The calabash from Lagos (Plate XLVI) is a good example of bold, simple, geometrical treatment. It is interesting to compare it with the very fine specimen from the Foulbe of the Cameroons (Plate XLIXb) and to notice the added interest given to the latter through the use of finely engraved pattern in white line on the dark ground. In contrast to the angular type of motif which is so commonly found throughout Africa, the Tiv, Ibo and Ibibio of Eastern Nigeria decorate their calabashes with gracefully flowing curvilinear designs said to be

49

D

derived from the patterns painted on the skins of these people. The Hausa in Northern Nigeria use curves in their designs and curvilinear patterns are also scorched on the gourds of the Kiga in Western Uganda (Plate LI).

In West Africa, in territories such as Liberia and Nigeria and most probably a number of others, leaf forms are used in the decoration of calabashes and other objects. They may flow in a fairly naturalistic way over the surface of the object decorated, or they may be arranged in a rather unimaginative stylized pattern.

When we turn to the use of representational forms of men, animals and other objects in calabash decoration we find much that really shows no intention whatever of design. The figures are scattered about the body in a haphazard fashion and all that can be said for it from a designer's point of view is that the spaces are all filled. Without some sort of geometrical framework within which to set the motifs it is impossible to think of the rounded body of the calabash in terms of pattern, and the craftsmen were probably only concerned with recording a story or setting down a number of symbols for some specific purpose, they were not interested in the decorative appearance of the work in the least. But in the work of some tribes, notably those in Dahomey, bands of geometrical motifs such as bars, zigzags, triangles, checks and lozenges often divide other bands or panels which are filled with representational motifs and these designs are extraordinarily successful. The contrast between the different elements is satisfying as a start, while the actual design of the panels with their grotesque, slightly heraldic, stylized animals is excellent.

CARVED COCONUTS
(Plate LIII)

Some well-carved coconut shells from Benin may be found in the British Museum. We reproduce one here in order to show the very different technical treatment from that used in calabash decoration. The coconut shell is carved in low relief, the background being cut away and the raised portions of the motifs being finished with a slightly rounded surface. The technique resembles that of Nigerian wood carving and the work is far more naturalistic than most calabash design. This particular specimen is crudely cut and dates back to the end of the last century.

References:
    Griaule et Dieterlen, 'Calebasses Dahoméennes', *Journal de la Société des Africanistes.* 1935. Tome. V. Fasc. II.
    Hambly, *Culture Areas of Nigeria.* Chicago, 1935.

CALABASH PATTERNS

Murray, 'The Decoration of Calabashes by the Tiv'. *Nigeria* 36. 1951.
Herskovits, *Dahomey*. New York, 1938, Vol. II, pp. 344 et seq.
Lindblom, *The Akamba in British East Africa*. Uppsala, 1920.
Pitt Rivers, *Antique Works of Art from Benin*. 1900.

CHAPTER X

Decoration on Wood

TEXTURE PATTERN
(*Plates LIV, LV, LVI*)

The richness of surface texture shown in African carving is discussed in Chapter XIV. It is perhaps a more noticeable pattern element in wood carving than in work in any other material. But a general desire to cover the surface is by no means all that may be found in these carved patterns. In even the most simple there is usually some division between plain and decorated surfaces, the areas being broken up into bands or masses, and a conscious relationship between the pattern and the form of the object is nearly always shown. In Plate LV the central mass of the bodies of the two powder kegs has been fitly emphasized by the curving forms cutting across the surface pattern, while the stool from Mashonaland is a beautiful example of the relationship of pattern and form. Finally in the wooden vessel from the Kasai shown in Plate LVI we have a magnificent example of African pattern at its best. The flowing mass of the interlacing scrolls on the body, balanced by the well proportioned and skilfully carved bands of pattern covering the neck, combine to make a most impressive yet simple design.

REPRESENTATIONAL FORMS IN CARVED DESIGN
(*Plates LVII, LVIII, LIX*)

Wood, especially the hard woods, is a material which has a fair degree of permanence;

52

it is also one which can be worked to a high finish by a skilled craftsman and it allows considerable freedom in the carrying out of all kinds of design. It is, in fact, the best material available over the greater part of the area which we are studying, and so it is natural that a very large number of the objects used in the religious cults and in court ceremonial as well as in architectural features should be made of it. So, owing to the purposes for which these things are carved, another element enters into their decoration, the use of representational forms and of symbolism. One of the most beautiful pieces of African carving, illustrated here, is the wooden door decorated with fishes by the Baule tribe of the Ivory Coast, a people well known for their aesthetic sense. The fish are naturalistic, but have been placed in relationship to the shape of the panel with all the skill of a Chinese painting. Their broken textured surface catches the light and stands out in contrast to the ground with its slight suggestion of waves. The total effect of the work is most lovely. Door panels and mouldings offer great scope for carving, and the Yoruba of Nigeria take full advantage of it. The panels are often crowded with tiers of detailed carving depicting processions of chiefs or European officials on the march, or scenes which doubtless have some historic or traditional significance. They often show a nice sense of humour in their portrayal, not unlike the misericords and corbels in our medieval churches. In some, which are less crowded, a sense of pattern over-rides the subject matter; in all, the enjoyment of texture pattern creeps in, in the decoration of figures and trappings, or in the bands which divide one portion of the design from another. The moulding of a door frame from Zanzibar is illustrated here to contrast with Yoruba work. It cannot be called fully African, for it is the work of Swahili craftsmen in the Arab tradition. Yet as Arabic influence is strong in the coastal region of East Africa, it is interesting to note the very different characteristics of the flowing curves and plant forms which are used.

Black and White Wood Carving
(Plates LX, LXIa, LXIIa)

An effective technique which has been elaborated in East Africa and possibly in other parts also, is one which is very reminiscent of that used in calabash decoration. The object to be decorated is first stained black all over, and the pattern is then either engraved in white on the black ground, or left standing in black against the ground which has been cleared away to a depth of about one eighth of an inch and left white. Various types of ladles and vessels are found worked in this technique on the coast, while the Lacustrine Bantu tribes of Uganda and the Eastern Congo use it for arrow quivers, as well as for milk vessels and containers of different kinds.

PAINTED PATTERN ON WOOD
(*Plate LXII*)

Those tribes which use wooden shields in East Africa often decorate them as they would shields of hide, with painted pattern which almost certainly has a symbolic significance, in chalk and coloured earths; the resulting irregular geometrical design is most attractive. In a similar way pattern is painted on wooden objects such as masks, canoe prows, and vessels of various kinds in many parts of the continent.

SCORCHED PATTERN ON WOOD
(*Plate LXIb*)

Pattern is also found burnt on wood, in a technique which combines the scorching of large areas with the engraving of lines with a hot iron point. The Ibibio fans illustrated here are an example of this kind of work.

DECORATION OF WOOD WITH EXTRANEOUS MATERIAL
(*Plate LXIII*)

Finally there is the decoration of wood by the application of some foreign material. This takes many forms; sometimes strips of metal sheeting are applied, or beads or metal studs may be hammered into the wood. Perhaps one of the most interesting ways in which the technique was developed is that which was used by the Kamba of Kenya in the decoration of their stools. The old craftsmen first wound a fine spiral of brass or copper wire by twisting it closely round a piece of metal like a fine knitting needle. This was then slipped off and hammered horizontally into the surface of the wood which had previously been softened by soaking in oil.

CHAPTER XI

Ornamental Ivory Carving

(Plates LXIV, LXV)

In Africa the carving of ivory has been carried out in a very similar fashion to that of wood. It is, of course, usually done on a smaller scale, but the fact that, unlike wood, it may be carved without regard to its grain, allows for very fine detail in pattern work.

In the West Coast, and especially in Nigeria amongst the Yoruba and the Bini, various decorative objects were carved from sections of tusks; these included armlets, bowls, boxes and so on. They are carved in low relief, sometimes to a considerable depth, or, as in the case of the bracelet shown in Plate LXIVa, the work is carved in two separate layers interlocking as a Chinese puzzle does, with the design of the outer layer cut right through. In this example the inner layer is decorated with a pattern of finely drilled holes, and the small bird seen in the centre at the base is carved on the inner layer, being one of the points which prevent the two portions of the bracelet from coming apart. In this specimen and also in the Benin tusk in Plate LXVb the very great freedom of design which we also find in Yoruba wood carving is clearly to be seen, nevertheless it is well balanced and consistent in the planning of its masses. The bowl in Plate LXIVb is much more disciplined in its arrangement, as the figures are spaced at a regular distance round the sides. This piece is most interesting in the way in which the figure on the right has been freed from the normal frontal position without breaking the rhythm of the whole.

The two illustrations of the pitcher in Plate LXVc and d are most delightful. Here bands of interlacing pattern separate the panels on which animal motifs are carved.

In these we find many symbols and stylized animals which we associate with the work of Benin; the two-headed bird, the mud fish and the highly stylized head of an elephant seen from above, with a double trunk ending in hands holding leafed branches. Other representations such as the charming little antelope grazing on the branch of leaves are much more naturalistic.

The freely flowing pattern typical of the carving on Benin tusks has already been mentioned. The general effect of these carvings is very rich, but the more simplified treatment shown in the two tusks from Loango in the Lower Congo is more intelligible to the observer. In the first the figures are arranged in a spiral round the tusk, while in the second they are set in bands placed one above the other.

In many African ivory carvings, details such as the spots on animals, the eyes and tribal marking on human masks or faces, and stylized representations of hair, or suggestions of neck ornaments, are inlaid with strips or studs of iron, brass or copper.

CHAPTER XII

Decorative Metal Work

BEATEN METAL DESIGN
(*Plates LXVI, LXVII*)

There are two very different methods of decorating metal; one is by beating, the other by casting; of these the first is almost certainly the older. The term repoussé is often used for any kind of beaten metal work, but its correct definition is the raising of a pattern in relief by blows with a hammer and punches from the under side. A similar effect can be obtained by beating down the background from the front with the aid of punches. The term chasing is applied to the surface decoration of a piece of metal from the front. By producing an uneven texture in which parts of the surface are tilted to the light, or by using punches which themselves have a pattern element, many different textural qualities may be obtained, and this satisfies the African love of surface pattern. Herskovits[1] says that in Dahomey the most important part of the equipment of every brass worker was a set of iron dies which he made for himself. These were used to represent figures on cloth, patterns on animal pelts, hair, eyebrows and the like on human faces. From the context this would appear to refer to the *cire perdue* method of casting described below, but most authorities agree that there the decoration is done on the wax. Interesting variations in texture may be seen in the two brass medallions decorated by the repoussé method illustrated in Plate LXVI. In the Ashanti vessel shown in Plate LXVIIb the motifs are outlined with small holes punched into the brass and filled with white material. While beaten metal work is found in Ashanti, Dahomey and Southern Nigeria it is far more common further north, the chief centres being Bida and Kano.

[1] Herskovits, *Dahomey*. Vol. II, p. 356.

57

DECORATIVE METAL WORK

Cast Metal Design
(*Plates LXVII, LXVIII, LXIX*)

Casting in West Africa is done by the *cire perdue* method. Briefly this technique consists in first making a rough model in clay of the object to be cast. This is then covered with a carefully modelled layer of wax in which every detail is exactly finished. The wax layer is constantly smoothed with a hot knife, and incised lines cut where necessary, while finely drawn threads of wax are attached to form raised lines. When the wax shell is complete and has hardened, it, in its turn, is encased in a clay mould, a duct being left through which the molten metal can be run in as the wax is absorbed into the clay when heated. This is done in a primitive charcoal-burning furnace. When the mould is broken and the metal cast removed the work is practically finished, for all the surface decoration which has the appearance of chasing was modelled in the wax. The technique of the casts made in Ashanti, Dahomey, and above all in Benin, in the past was of a very high quality indeed and has given rise to much discussion as to its origins.

It is not easy to define the right qualities for design in metal casting as the work is not done directly on the material in which it will finally appear. The motifs have to be modelled or incised in soft wax which has very different qualities from the hard metal in which they will eventually be reproduced, so that the normal technical criteria of fitness for material and the appropriate tools for working in that material are missing. The motifs used in Benin bronze casting are discussed fully in the appropriate chapter; it is sufficient to notice here that both human and animal forms, or, on the other hand, purely geometrical design can be equally well carried out by the process, and where natural forms are used they may be very realistic or highly stylized. The curious combination of a comparatively small number of accepted stylizations within a representational formula in Benin casts is possibly partly a reflection of the lack of structural guidance in this technique.

CHAPTER XIII

Pottery Design

While in metal work, wood carving, cloth making and printing the African craftsman has developed very considerable skills, and exercises his ingenuity in working out methods of casting, constructing looms, printing and dyeing cloth and so on, yet with few exceptions his achievement in the craft of pottery has never reached an outstanding level. The preparation of clay is very perfunctory, nothing is known of an indigenous wheel beyond a simple turntable, no kiln is used for firing, and the use of a vitreous glaze is not attempted. Nor has the African aspired to build any great variety of shapes for his pottery, or to put it to many different uses. True, he has used pots for various magico-religious purposes from burial urns to pots containing poison or magic potions or sacrificial libations, but on the whole his chief need has been for cooking pots, water pots and beer pots, all vessels which will be subjected to rough and dirty treatment, and so unlikely to develop in him a desire to expend time and labour on the carrying out of delicate design.

So the soil of Africa from north to south and from east to west is littered with ill-fired sherds on which elementary pattern forms are scratched or impressed with roulettes of wood or of plaited reed, or by stamps made of natural objects such as corn cobs, shells and seeds. All these are interesting and even excellent examples of decoration by primitive methods, but one is struck by the lack of development and the difficulty of finding outstanding specimens which show a real feeling for design in the craft.

The most widespread methods of decorating pots found in Africa are those of impressing or stamping a pattern, or of engraving or incising one with a pointed stick or piece of iron; both these are carried out before firing when the pot is leather hard. Moulded decoration is also found, and in some areas pots are coloured or burnished.

IMPRESSED PATTERN
(*Plate LXX*)

Accounts of pot making and descriptions of pots and sherds from every part of the continent suggest an enormous variety of natural and manufactured objects used to impress simple patterns on the clay. Such decoration may spread all over the pot, or it may be formally arranged in zones often bounded by incised lines. The whole surface of the pot may be smoothed with a small stone placed in a damp cloth, when slight pressure in working will cause small circular indentations which satisfy the desire for interesting texture. This characteristic love of texture pattern which we notice in every African craft will find an equal pleasure in closely covering the surface with sharply cut indentations made with the finger nail. The sharp edge of a shell may be used to imprint a series of small holes or circles on the clay, or, to quote another description: 'Compressed chevrons are made by rocking and advancing a mussel shell held in the right hand. The shell is rocked on its rounded edge and moved simultaneously in a direction at right angles to the plane of that edge.' Parallel tracks may be hollowed out by drawing the tips of the fingers across the surface of the pot, and a series of little indentations by regular pressure with a piece of corn cob. The custom of plaiting little reed roulettes which are rolled round the neck or body of the pot to form bands of pattern is very common; these plaited roulettes will make a dotted diagonal or herring-bone pattern according to the way in which they are knotted (Plate LXXd). Many patterns are also made with roulettes carved from small pieces of wood, while some tribes impress a pattern with an iron ring. Our three illustrations of impressed pattern show first a large water pot from Uganda which is a fine specimen of a pot decorated with bands of roulette pattern. Then comes one which is indented, probably with a finger nail, over the whole surface of a clay face, but which is carefully arranged to emphasize the form. Notice the central dividing vertical line above the nose, with the curves following the eyebrows on either side, and the tall rectangular panels suggesting the cheeks. Finally pattern confined within definite bands. Such pattern is, practically without exception, rolled round the body of a pot in a series of concentric horizontal circles, and the specimen here shown is most unusual in the way in which the pattern element has been arranged on the body.

60

POTTERY DESIGN
INCISED PATTERN
(*Plate LXXI*)

Equally common, as a method of pot decoration, is the incising or engraving of a pattern with a knife, pointed stick, nail, shell, piece of wood or gourd with a serrated edge like a comb, or any other pointed object. The two styles of impressing and incising are often used together. Again this type of work is often crude and haphazard, but many well-shaped pots have bands of incised pattern emphasizing the changing silhouette, or are ornamented with geometrical shapes such as bands, triangles, lozenges, and so on, filled in with parallel lines or cross-hatching. These incised surfaces may be further emphasized by the filling-in of the engraved lines with white kaolin or crushed shell or taluka powder, or they may be arranged as matt surface forms surrounded by smooth areas burnished to a high gloss. Of this burnishing more will be said later.

MOULDED PATTERN
(*Plates LXXII, LXXIII, LXXIVa, b*)

A number of ceremonial bowls and terra cotta boxes may be found both on the West Coast and in the Congo which have human figures or animals modelled in high relief on the lids or body of the pot. Some of these fall rather into the class of small free-standing sculpture than that of decorative design. In Plate LXXIIb and c, are shown Ashanti pots in which the moulded element takes its place as part of the general design; while LXXIIa is a good example of the stylized treatment of figures in both Cameroon pottery and wood carving, when the arms and legs form a kind of lattice work round the body of the object.

Plate LXXIII shows various types of curvilinear moulded pattern, while Plate LXXIVa and b are two examples of patterns made by attaching pellets of clay to the body of the pot.

COLOURED OR BURNISHED DECORATION
(*Plate LXXIVc, d*)

Next we come to the use of burnishing and the coating of pots with coloured vegetable or mineral matter. These processes may have either an aesthetic or a practical purpose; in some cases it is obvious that the desire to enhance or entirely conceal the

61

dreary colour which is taken on by so much fired clay is the main reason for what is done; in others the practical advantage of rendering the vessel less porous is the chief motive. A very interesting account of colouring methods practised in the Belgian Congo is given in the *Annales du Musée du Congo Belge*.[1] The subject is divided into three sections; first the application of vegetable matter giving a matt surface, secondly of vegetable matter giving a glossy finish, and finally the use of mineral substances. For the first palm oil or other plant juices are applied after firing and are often put on as an even coating over the whole pot: in this case any variation in colour is fortuitous. At other times more purposeful decoration is obtained by painting or splashing on the colouring matter. In this way the people of the Lower and Middle Congo produce pottery with the appearance of marbling or the grain of wood; or, in other cases, the pot may look much like the map of an ocean filled with coloured islands and archipelagoes.

In some areas rudimentary decoration in the form of lines and geometrical motifs is painted on with the finger or a piece of stick, while in the Lower Congo Coast region motifs depicted in this way show signs of a zoomorphic origin.

Various gums and resins are used to give a glossy finish to the pots. Sometimes these are painted over the whole vessel, in others a more interesting effect is produced by leaving some areas with a matt surface, which, as has been said, is often filled in with an incised or impressed pattern.

Mineral matter, usually in the form of graphite, or of an iron ochre mixed with oil, is sometimes rubbed into a leather-hard pot and burnished before firing in many parts of Africa. As hard rubbing is required to produce a highly polished surface, no portion of the pot with incised or impressed pattern will stand up to the treatment; so that the decoration of the pot will consist in the arrangement of plain burnished masses contrasted with the decorated areas having a matt surface.

ENCASED POTS
(Plate LXXVIa, b)

An interesting little pot is shown in Plate LXXVIa. The neck and base are completely hidden in a woven fibre covering, which is continued as a handle. The body of the pot has a decorative binding of strips of cane, so that all that can be seen of the pot itself is a series of panels framed by the cane and decorated with roulette pattern. The planning of the total decorative effect, to include both the elements of the actual body

---

[1] 'Céramiques et Nattes'. *Annales du Musée du Congo Belge*. Series III. Tome II.

of the pot itself and also the extraneous casing, is notable. The second encased pot has an equally pleasing appearance, but in this example the entire surface of the pot is covered.

Cow Dung Bowls
(*Plate LXXVIc, d*)

Two other vessels are also illustrated, though neither, strictly speaking, comes into our geographical field of study, nor into a rigidly defined classification of pottery; but both should be of real interest to a student of African pottery design. These are a couple of bowls made of cow dung from the Nuba Hills in Kordofan. The pattern is painted in white and red earth, giving a rich and well designed effect which might well be adopted by the makers of pottery proper.

References:

Thomas, 'Pottery Making of the Ibo Speaking People', *Man.*, 1910, p. 53.

Schofield, *Primitive Pottery: An Introduction to South African Ceramics.* Handbook Series No. 3 of the South African Archaeological Society. Capetown, 1948.

Nicholson, 'Bida Pottery', *Man.*, 1929. 34.

Griaule et Lebeuf, 'Fouilles dans la région du Tchad'. *Journal de la Société des Africanistes.* Tome XVIII. Fasc. I.

Trowell, 'Some Royal Craftsmen of Buganda', *Uganda Journal.* Vol. VIII.

CHAPTER XIV

Motifs in African Design

Although technological considerations play such an important part in defining the possibilities and limitations of decorative design that it has seemed best first to approach our subject from this angle, it is equally necessary and possibly more interesting to study the motifs used, regardless of the way in which they are worked out. So now, having thought through the implications of method, let us briefly examine the matter of African pattern.

It would seem obvious to divide it into three types; those of textural, representational and geometrical forms. At once we realize that there are no hard and fast boundaries to these groups, and that the one slides imperceptibly into the other; while all the time we are working along the edge of the impenetrable jungle of symbolism. Texture pattern, however, seems to be the most easy to define. Until comparatively recently any European study of design dealt only with specific pattern motifs; it might be classical ornament or floral pattern but, as such titles as 'The Anatomy of Pattern', 'The Grammar of Japanese Ornament and Design', or 'The Dictionary of Decorative Art' will show, decoration was conceived of as the ornamentation of an object with representational, even if stylized, forms; or at least with strictly geometrical designs. Students in schools of art studied natural forms, especially those of plants, they manipulated and twisted the plant as a whole, they dissected the flower and took cross sections of the seed capsules, and selected and re-grouped their results until they covered a surface with symmetrical clusters of flowers or fruits. It is only lately that our great interest in texture has developed. Perhaps it is due to our own preoccupation with its decorative possibilities that we notice the presence of texture pattern in African design; nevertheless, once we have noticed it, its predominance is most striking.

MOTIFS IN AFRICAN DESIGN

In making a study of any collection of African work we cannot fail to be struck by the extraordinary richness of surface texture which is conveyed through the use of small patterns over large areas. It is perhaps most noticeable in work on the harder materials, pottery, metal, wood, calabash and ivory, although it would seem highly probable that the attention of African craftsmen was drawn in the first place to the decorative interest of such pattern as it arose through the technical process of fibre weaving. For even the most simple weaves found in basketry, matting and woven textiles have satisfying textural effects; and a very slight variation in the order of work, or the substitution of another weight of fibre or a different colour in places, will alter the appearance of the whole thing. Such a simple form of ornamentation conveys no abstruse symbolic meaning, it is meant to be enjoyed for its own sake alone (Plates XVb, c, XIXc, XXXIIIc).

Similarly the handling of pottery before firing will give rise to natural forms of pattern; it may be slight irregularities of texture due to the pressure of the tool used in smoothing the vessel, or a groove caused accidentally by the finger tips, or a mottled effect obtained unintentionally when covering the pot with a coat of vegetable matter to render it less porous. It is often said that the idea of impressing a pattern on to the soft clay came from the custom of lining baskets with clay and then burning off the fibre covering in the fire, thus leaving a terra cotta vessel with a basketry pattern on the outer side as though imprinted from a mould; whether this be true or not it is a short step to develop from pattern effects accidentally achieved in the making of a pot, to ones deliberately formed by impressing it with a finger nail, corn cob, or roulette. The engraving of pattern with a sharp point was developed simultaneously, and in their simplest forms such patterns may cover the whole surface of a pot (Plate LXXa, b).

More formal use will be made of such decoration through the alternation of zones of plain and textured surface, or of areas in which the textural patterns differ; as the work reaches this stage it is not always easy to differentiate between what we have called texture pattern and geometrical design (Plates LXXc, LXXIa, b).

When we turn to decoration carried out on such materials as wood and calabash, ivory or metal, we find that the tendency is either to use texture pattern to fill up the background against which the forms of men or animals are set or to keep the background plain and to cover the forms with fine patternwork often representing the texture of clothing, animal pelts, or fish scales. At other times these texture patterns on the figures are entirely unrepresentational and merely serve to give continuity across the areas in relief; when background pattern is used it, in its turn, may be either completely non-representational or may consist of simple repetitions of floral or other symbols

MOTIFS IN AFRICAN DESIGN

(Plates XLIXb, LIV, LVb, c, LVI, LVIIIb, LX, LXI, LXIV, LXVI, LXVIII, LXIX).

REPRESENTATIONAL FORMS

Proverbs, allegories, adages and wise words form a backcloth to African thought, and both the myths and legends of the tribal past and the prestige of the reigning chief or king are summed up in aphorisms or visual symbols. It is this which gives us the key to much of African design. Most of the traditional pattern motifs of the various tribes had originally a symbolic or allegorical meaning. The initial intention of the craftsman when he carved or embroidered some animal or hieroglyphic was not to decorate his handiwork, not to depict some animal of which he was particularly fond, not even to make some mighty fetish, but to record a pictorial statement of an idea. The meaning may have been clearly stated or abstruse, it may have been kept secret to a privileged few, and in course of time it may have been completely forgotten, nevertheless it was the original reason for his act, we might call it the motive of the motif.

In this the African does not stand alone. To quote Ethel Lewis,[1] who has written an interesting account of the use of symbolism in textile design throughout the world, from the early civilizations through Europe, the Orient, and South America, down the ages, 'The patterns created in Egypt were distinctive though evolved in the same way as practically all early design. They grew from the need to represent in some way certain religious symbols. Carlyle has described decoration as the first spiritual want of man. And from that early need right down to our own times ornament has served as an expression of life. There is a story in each pattern, for every bit of ornamentation is a symbol of something, and each symbol is a record of history or experience.'

In monographs on specific examples of African design, such as wall painting or textile printing, one reads that modern work in the particular medium is representational while older examples are usually of geometrical pattern. It is very possible that in its early stages the older work was also as representational as its maker wished it to be, but that through successive copying and stylization it became reduced to a geometrical symbol accepted and understood in its own day but unintelligible to us now. The unravelling of the meaning of these African symbols is a task which we must leave to the ethnologist; and it can only be based upon many years of study sympathetically undertaken in all parts of the continent. Even so, it is doubtful whether at this late stage we shall ever be absolutely sure of all the interpretations which we are given. For even such reputable authorities as the late R. S. Rattray, and Eva Meyerowitz, may come to

[1] Ethel Lewis, *The Romance of Textiles*. New York, 1953, p. 5.

66

different conclusions as to the ideas behind such motifs as those cut on the stamps used in printing Adinkira cloths. Rattray[1] believed that the Ashanti borrowed these motifs from the amulet symbols used by the Arabs from the north but renamed them giving them a local historical, magical or allegorical significance; while Mrs. Meyerowitz[2] fits a number of them into an elaborate scheme representing different aspects of the Creator, adding a warning that although the real meaning of the symbols is known in secret by the craftsmen and their descendants, they are used freely and often other secondary meanings have been substituted for the original ones. Both these authorities may well be right, and one set of meanings may be secondary to the other, but it is obvious that for the amateur who has not studied the subject on the spot, any attempt at exegesis is fraught with danger. Happily the designer's field of study is something other than the interpretation of the content of these symbols, for his sphere is the appreciation of the form in which they are depicted, and although his task will be made both lighter and more interesting by all that he can learn of the meaning which they hold, yet their initial value for him lies in their visual appearance.

In the decorative art of certain parts of Africa, notably that of the West Coast Kingdoms, we find much use is made of both human and animal forms, which are depicted in every degree of stylization from almost naturalistic representation to the near-abstract symbol. No purpose would be served by illustrating these forms and symbols in a haphazard fashion from many different tribes, and it has seemed best to concentrate on the work of one or two regions where their use has been most highly developed. In many cases it would seem that these motifs are used to portray the glory and honour of the tribe as personified in the ancestors or living rulers. These rulers are often represented in the form of animals to whom various approved characteristics are traditionally ascribed. The art of Dahomey affords a most excellent example of such a form of analogous representation. There, in bas-reliefs of clay which decorated the palace walls, and in appliqué hangings, umbrellas and banners, as well as in work in other media, the great deeds and victories of the kings, their glory and their powers are all set out in symbolic form. During the course of his reign each king acquired a series of praise-names which referred to an existing proverb or one which was coined to describe some incident in his life. Thus when the king Gezo supplanted his brother Adaza it was said: 'The horse carries a bit which the ox cannot carry,' and this found visual expression in the head of a horse wearing a bit of iron. Other such representations are a draped buffalo—'The cloth which falls on the buffalo's neck cannot be

---

[1] Rattray, *Religion and Art in Ashanti*. Chap. XXV.
[2] Meyerowitz, *The Sacred State of the Akan*. Chap. VI.

E*

removed'; a fish—'the fish who refuses the net cannot re-enter'; or a young lion who spreads terror abroad directly he has cut his teeth. Sharks, crocodiles, dogs, panthers, hornbills, toads, pythons and many other animals, birds and reptiles are used in a similar way.[1] In many cases one motif cannot carry the whole content of a proverb or a praise-name or represent the victorious action of the king in battle, so a whole panel is constructed of motifs which must be read together to understand the analogy (Plates V, XXV).

Another form of ornamental art carried out by the Dahomean people is that of calabash decoration. According to Herskovits[2] one type of elaborately decorated calabash is sent, with a suitable gift inside it, by a young man to the girl who has won his heart. Just as the royal bas-reliefs and appliqué cloths depict in allegorical form the glories of the kings, so these love tokens convey through groups of symbols, which represent well-known aphorisms, the state of the young man's affections. In these birds, animals, fish and serpents are represented together with numerous other objects such as stylized human faces or hands, palm trees, pots and knives.

The Dahomean calabashes illustrated in this book come from the collection made by the Dakar–Djibouti mission in 1931, and have been described by Griaule and Dieterlen[3] who state that they are to be found in huts and market places, but give no explanation of any allegorical meaning. Both these writers and Herskovits say that the use of the whole human figure in such calabash decoration is extremely rare (Plates XLVII, XLVIII, XLIXa).

Except for variations which are naturally due to the use of different materials and techniques, the treatment of natural forms in all these branches of decorative design in Dahomey is somewhat similar. The animals, and human figures where they are used, are first thought of as a solid silhouette, portrayed as naturalistically as is consistent with the artists' purpose. That is to say, in the bas-reliefs and appliqué cloths where the aim is chiefly descriptive and pictorial, liberties will be taken with the total size and relative proportions or stance of a figure or animal if by so doing the conception of grandeur or power can be enhanced. In calabash decoration, on the other hand, where the feeling for ornamentation is stronger, the animal forms may be elongated, compressed or distorted to fit them within the shape of the panel allotted to them, and the surface will be

[1] Mercier, *Etudes Dahoméennes*. V.
See also Herskovits, *Dahomey*. Vol. II, pp. 328 et seq.
Waterlot, *Les Bas-Reliefs des Bâtiments Royaux d'Abomey*.
[2] Herskovits, *Dahomey*. Vol. II, pp. 344 et seq.
[3] Griaule et Dieterlen, 'Calebasses Dahoméennes', *Journal de la Société des Africanistes*. Tome V. Fasc. II. 1935.

broken by bold pattern. Yet in both classes of work the subjects are kept as separate compact masses silhouetted against a plain ground.

When we turn to consider the use of animal and human forms in the art of Benin we find a completely different approach. Let us first try to make a composite picture of the actual appearance of the king and his courtiers from the records of the early explorers, for comparison with their representation in metal, ivory or wood.

'On entering the apartment I perceived his majesty; a fine, stout, handsome man, with somewhat of kingly dignity about him. . . . Several chiefs, in full dress, surrounded him, beside a bodyguard on either side, with drawn swords. On my approaching him, he held out his arm, which the three young princes stepped forward and supported.' Fawckner. 1825.

'Lastly came the king, supported by two men, who led him to the wooden bench upon which a mat had been placed, disposed his loin cloth and held both his arms.' Burton. 1862.

'He [the Captain of war] was standing in the upper alcove with two attendants gingerly supporting his coral and iron braceletted arms, which hung down loose and away from his sides. The general effect of his attitude upon a newcomer was that of a fainting man being caught in the act of falling . . .' Burton. 1862.

'The rich among them wear, first, a white calico or cotton cloth, about one yard long and half as broad, which serves them as drawers; over that they wear a finer white cotton dress, that is commonly about sixteen or twenty yards long, which they wind very neatly round the middle, casting over it a scarf of about a yard long, and two spans broad, the end of which is adorned with fringe or lace . . . the upper part of the body is generally naked.' Nyendael. 1704.

'The nobles were bare to the waist, save for a few strings of coral beads of no large size. Below the waist they wore very full petticoats.' Punch. 1889.

'He [a court official] was curiously habited, wearing a sort of short petticoat from the waist down to the knees, composed of a cloth very much valued by them, resembling our white bunting. This encircled his loins, and was set off like an ancient dame's hooped petticoat; the upper part of the body was naked, as well as the legs and feet.' Fawckner. 1825.

'He [the king] wore a net shirt of which each knot was furnished with a coral bead; it weighed more than twenty pounds.' Landolphe. 1787.

'The "Big men" had each his anklets and collar of coral, a very quaint decoration, composed of pieces of about one inch long, and so tight strung that it forms a stiff circle about a foot in diameter.' Burton. 1862.

'The chief's head was covered with coral beads, thickly strung on his black curly wool; his neck, ankles and wrists were also encircled with a profusion of strings of coral.' Fawckner. 1825. He also noted that besides the necklace which the king wore on state occasions he also wore a belt of coral.

[a noble] 'held an *ebere* in his hand which he kept twisting round.' Punch. 1889.

'In his hand he held a fan made of leather . . . to keep off the flies and protect him from the rays of the sun.' Fawckner. 1825.

'It was always quite the thing for an African potentate to have at least two slaves fanning him.' Punch. 1889.

When, after having read these descriptions,[1] we look at the bronze plaques representing the king and his courtiers, either with naked torsos and 'full petticoats' or with a net shirt covered with coral beads; with their chokers, armlets and anklets of coral; their fans and *eberes*: and note how the attendants are grouped about the king, fanning him and protecting him from the sun and supporting his arms, we tend to say that this is pure naturalism, with no attempt at symbol or stylization. Not only is every detail of dress meticulously recorded, but many figures break away from the position of frontality and present a side or even three-quarter view. Other works representing Portuguese, or African scenes apart from court life and people, are even more naturalistic. Yet in the midst of all this seemingly straightforward representation we perceive a strong symbolic element which is fitted in with no sense of incongruity. On some carved tusks the king's legs become a pair of mud fish curving upwards and ending in a head, on a plaque two snakes sprout from the nostrils of a human face, or the trunk of an elephant ends in a human hand. It is this dualism of the lifelike and the fantastic at one and the same time which is so remarkable in Benin art.

The mud fish element, together with serpents, double-headed birds, and other fantasies, must be fitted into Benin mythology; for they doubtless make reference to the gods; but there is one interesting stylization which has no supernatural meaning and has come about from a straightforward kind of shorthand. In the bronze plaque representing a Portuguese, Plate LXIXb, the head is shown in a most naturalistic way. The helmet has a broad brim, the hair is long, and combed downwards and outwards straight to the shoulders, and the beard ends in three points. In the version on

[1] All quoted Ling Roth. *Great Benin.*

the small plaque, Plate LXVId, some stylization has already taken place. The helmet has become a jaunty hat with up-turned brim, and the hair ends in a large curl on each side of the face. Finally the very formalized motif which is repeated over the blade of the *ebere* in the same plate has taken the elements of hat, hair and beard and turned them into a symbol for a Portuguese which we recognize only through having studied it in the earlier stages. This is an excellent example of the reduction of a natural representation to a completely stylized symbol (Plates LIXd, LXIV, LXVb, c, d, LXVI, LXVIII, LXIX).

In both the Dahomey and Benin work representational forms make up the chief part of the design. They are usually arranged in panels framed by borders of geometrical pattern and must be considered to have a literary as well as a decorative value. A some-what similar approach can be seen in the wall paintings, wood and ivory carvings, printed cloths and metal work of other tribes illustrated here (Plates I, II, III, LII, LVIII, LXVa).

Then comes work in which human and animal forms seem to hold a less significant place, and are used either as symbolism or for their purely decorative value as a part of a general scheme of design (Plates XIc, XVIIc, XXXIIId, XLIII).

Still further developed are the motifs which obviously have some symbolic meaning or reference, but which have to be interpreted to all but the initiated (Plates VIIIb, XIII, XIVb, XXVIII); and finally those forms which to all intents and purposes appear to be purely geometrical. In all representational design animal motifs seem to predominate, followed by representations of the human figure and certain types of objects, while the use of floral forms is remarkable for its absence.

SYMBOLIC AND GEOMETRICAL PATTERN

In considering the use of geometrical and symbolic pattern in African design it has seemed more profitable to restrict ourselves in a similar way to that taken when study-ing human and animal motifs, and then to consider one or two areas where it is possible to find a fairly full account of its use. For this purpose we have taken the work of the Bushongo and neighbouring peoples of the central Congo.

When we come to analyse the pattern motifs in Congo embroidery and wood carving we find that it is not possible to work out any clear scheme of classification. Torday and Joyce, who made a very thorough study of the subject in 1910, said that the people themselves found it very difficult to name even well-known patterns with which they were quite familiar, and when they tried to do so, many arguments occurred, so that the matter had to be referred to expert craftsmen. The cause of their difficulties was that, unlike the European who regarded a pattern as a whole, the Bushongo appeared to break it up into a number of different elements and arbitrarily chose one to give the name to the whole design. This resulted in patterns which to us would appear to be quite dissimilar, being called by the same name; while others which we would consider similar were regarded as having no relationship at all to one another. Again, identical patterns were used in wood carving, and in tattooing the body, and in embroidery; and as both tattooing and carving were the prerogative of the men while embroidery was that of the women, the different sexes were likely to pick out different motifs as the key to the whole design, with the result that the same pattern would receive different names from each.

A very large number of the motifs are basically built up from interlacing pattern, and Torday and Joyce have attempted to elaborate the various stages through which they have passed, from the genuine technical necessities of weaving to their present form in embroidery and wood carving. The twisting and knotting of string in net making might seem an even more obvious source for certain of these patterns, especially as many riverine tribes use, and presumably make, large fishing nets. Another probability is that some are derived from the drawing games played by the young children of the tribe. These games consisted in drawing certain traditional designs in the sand in such a way that the whole figure was drawn in one continuous line without raising the finger from the surface. When the sand was damp the first part of the line would be eradicated in places by the crossing of later portions. This method of drawing would not result in a regular under-over, over-under, form of interlacing but would at least suggest the possibilities of such treatment. Similar games of pattern drawing with one continuous line are reported from Angola.

Other motifs in Congo pattern are derived from triangles, lozenges and so on; while from the names given to others it seems likely that they derive from old stylized representations or symbols. Such are 'Kanya's fingers', 'the drum of Mikope', 'Buffalo horns', 'the stones', 'the back of the wild cat', and 'the Chameleon's footsteps'.

Let us try to identify some of the patterns shown in our illustrations. Perhaps the most commonly used motif in Bushongo design is *Mbolo*. This would appear to be derived from the straightforward interlacing, as in weaving, of two pairs of strands, the four points of crossing being the essential element of the motif. The ends of the strands may be joined together in various ways, forming complete motifs such as are found carved on the lids of boxes, or embroidered in small sections of the pile cloths; or alternatively the interlacing may be continued as a long band of pattern on both carving and embroidery. The first type of *Mbolo* can be seen in three panels of Plate

XXb (1), on the right hand side of XXb (2), and in Plate XXb (3), where it forms pairs of oval links as a small central motif in parts of the pattern. The second type is shown in Plates XXb (6), XXI (8), and LIVd. Other variations will bear names, such as *Mongo*, the knee illustrated in Plate XXI (9), where the lighter portion shows the pattern element. *Mbolo* is one of the named motifs used in the sand-drawing games played by children which have been mentioned above.

Another motif, or set of allied motifs, arises from the twisting together of two cords. In this simple form it is called *Namba*, the knot. But the identification of this motif is just as confused as that of *Mbolo*. Apparently the cords in *Namba* should lie close together, forming no open loops between the twists; if they do leave such gaps the pattern becomes *Nyinga*, the smoke; and if the twists are accentuated into a very sharp angle and the gaps are also noticeable it is known as *Nemo Kanya*, the fingers of Kanya. Unfortunately, we have no example of this simple form of *Namba* in our illustrations, although the curvilinear interlacing pattern on the body of the pot shown in Plate LVI would appear to fit the description. According to Torday and Joyce, however, this pattern when used in wood carving by the men is called *Buina* because of its resemblance to the curved blade of the knife which they use. The angular *Nemo Kanya* is shown in Plate XXI (11).

A further state of confusion is reached when we learn that virtually any other form of twisted cord might be called *Namba*. The continuous single loop twisted on itself which we see in Plate XXb (3 and 5), is *Namba*, and so are many representations of twists and knots which would definitely seem to be based on netting and which are carved freely in the decoration of wooden vessels.

It would be fruitless in a book on general design to attempt to unravel further complications in the pattern names of the Bushongo; our purpose is served once we have realized that there is no easy road to recognition or understanding. Certain patterns, however, are easy to pick out although the connection between their name and their appearance may be quite obscure. Such are *Mosala Baba*, the feathers of Baba, shown in XXb (4), and XXI (12), where a group of four right-angled triangles are fitted together at their apex. The chevron pattern is known as *Mamanye*, the stones, and is illustrated here in Plate XXI (3).

Finally come a few in which the allusion conveyed by the name is fairly clear. Of these the most obvious is *Mayulu*, the tortoise in Plate XXI (5), suggested by the tortoise-shell; *Ikunji*, the eye, Plate XXI (7), where the triangular space suggests the subject; and perhaps *Lori Yongolo*, the feet of the chameleon, Plate XXI (10), conveys rather well the creeping gait of its namesake. It is interesting to compare these names with those of a very similar type given to the plaited matting patterns of Zanzibar

(Plate XIII); in both cases the pattern would appear to have first been made for technical or decorative reasons and named afterwards when the resemblance to an object was noticed.[1]

Although written over sixty years ago, Haddon's *Evolution in Art* must still be considered one of the classics on this subject, and nothing could sum up better the points which we have put forward here than two widely separated quotations from that book: 'By carefully studying a number of designs we find, providing the series is sufficiently extensive, that a complex, or even an apparently simple pattern, is the result of a long series of variations from a quite dissimilar original. The latter may in very many cases be proved to be a direct copy or representation of a natural or artificial object. From this it is clear that a large number of patterns can be shown to be natural developments from a realistic representation of an actual object, and not to be a mental creation on the part of the artist. There are certain styles of ornamentation which, at all events in particular cases, may very well be original, taking that word in its ordinary sense; such, for example, as zigzag lines, cross-hatching and so forth. The mere toying with any implement which could make a mark on any surface might suggest the simplest ornamentation to the most savage mind. This may or may not have been the case, and it is entirely beyond proof either way, and therefore we must not press our analogy too far. It is, however, surprising, and it is certainly very significant, that the origin of so many designs can now be determined, although they are of unknown age.' 'In studies such as these, the investigator should refrain from theorizing as far as possible; it is a dangerous game, for more than one can play at it, and the explanation is as likely to be wrong as right. The most satisfactory plan is to gather together as much material as possible, and it will generally be found that the objects tell their own tale, and all that has to be done is to record it.'

[1] The whole of this section is based on the very full account given in 'Notes ethnographiques sur les peuples communément appelés Bakuba'. *Annales du Musée du Congo Belge*. Series III. Tome II. Fasc. I, pp. 215 et seq.

Index

(Arabic numerals are page references, Roman numerals plate references.)

75

INDEX

76

INDEX

INDEX

Plates

PLATE I. WALL DECORATION

Top: Part of the painting of the Oku wall. By an Ibibio artist for Ibo tribe. Provenance, Bende Division. Nigeria.—This work was done by an Ibibio artist hired by the local people. It consists of 'a wall covered with gaily-coloured pictures showing in panels policemen, girls, cows, snakes, dancers and so on, singly and in groups. The borders between the pictures are often filled with patterns based on those of cloths. Each picture is complete in itself and there is no continuous story running through all, nor is there any connection in them with the customs or history of Olokoro; they are Ibibio. It is noticeable that everything is painted separately in a side view without one thing covering another. Several of the panels are very effective decorations in themselves, especially those with an arched decoration above the figures, and that of two dancers with long-tasselled headdresses and tails . . . The paintings are done on a clear white background—the old method used by the women of Bende was to polish the wall until the bits of mica in the mud made it shine with a golden sheen, and then to paint it with ochres, chalk and soot. At Okwu red and yellow ochres, pink clay, black and brown from a sap, and European manufactured washing blue and green powder paint have been used without any medium, such as gum, to make the pigments adhere'. See *Nigeria* 27. 1947.

Bottom: Wall painting from the Inre Court House. Provenance, Nr. Awka. Ibo tribe. Nigeria.—Here the stylized drawing of the men in canoes is very interesting. At the base three canoes are seen, viewed end-on; while above them the men are superimposed on what may be a bird's eye view of a canoe.

PLATE II. WALL DECORATION

Top: Wall painting. Southern Tanganyika.—A painting of a legendary chief, Kifutumo, from a collection of wall paintings of the *Buyeye* and *Bugoyangi,* secret societies of snake charmers in Southern Tanganyika.
See *Wall Paintings by Snake Charmers in Tanganyika.* Cory, London, 1953.

Bottom: Wall painting from an *Mbari* house. Near Aba. Nigeria.—These houses are filled with large mud figures and completely decorated with wall paintings. These consist chiefly of geometrical patterns, but also include panels containing stylized paintings of birds, animals and men, or highly intricate abstract pattern. The colours used are white, black, slate-grey, ochre, earth-red and washing-blue. This particular example with its patterning of dots and small circles gives the appearance of a mosaic. See *Nigeria* 49. 1956.

PLATE III. WALL DECORATION

Top: Wall painting. Probably Chokwe tribe. North East Angola.—A most comprehensive account of the wall paintings of some tribes of Angola is given by Redinha. (*Paredes Pintadas da Lunda*. Lisbon, 1953.) He tells of paintings made on the outer walls of huts, not by skilled craftsmen but by the ordinary people, for the sheer pleasure of decoration. Such work is of a very ephemeral character, and is technically amateurish; nevertheless it has vitality and a certain richness of colour and pattern. The subjects chosen are very varied, and consist of scenes from everyday life or legend. They are often schematic in their representation, and even when designs appear to be purely abstract, they prove to be crude attempts at topography. Amongst these people as elsewhere large integrated compositions are almost unknown, and the decoration consists in small isolated figures, or groups of figures or geometrical patterns.

Bottom: Wall painting. Provenance, Ekibondo Village. Niangara District. Bangba tribe. Uele. Belgian Congo.—In this small village, consisting only of some twenty huts, the old craft of wall painting has been revived by encouragement from the District Commissioner. This panel, representing a modern trend in wall decoration, is filled with fairly naturalistic snakes, fish, knives, and so on. Other similar ones show elephants, leopards and other animals, and men. All the elements, which are taken from well known stories, are placed side by side. As in most primitive art the figures are static, the body is presented frontally, the head and legs in profile. Each line or form is incised in outline on the wall and then filled in with colour. Colours used are ochre, white (kaolin) and black (charcoal, crushed with a certain leaf).

PLATE IV. WALL DECORATION

Top: Abstract wall painting. Provenance, Nyanza Province. Luo tribe. Kenya.—These were painted by a local craftsman on the walls of a schoolmaster's house between 1940 and 1945. Unfortunately no further information about them is now available. The designs are excellent.

Bottom: Symbolic wall painting. Provenance, Ekibondo Village. Niangara District. Bangba tribe. Uele. Belgian Congo.— This panel is a more ancient form of decoration than that shown in Plate III (*bottom*). It consists of abstract symbolic forms, which probably date right back to the religion of the Sudanic ancestors of this tribe. It is said to represent an old myth from the cult of the sun and the moon, the central rosette signifies the sun. The cross bands of pattern represent the moon, and the undulating lines joining them are 'the feet of the moon'. The sun is always present, but the moon walks with the rain and so must have feet. The motifs are welded together into a rich geometrical pattern.

PLATE V. WALL DECORATION

Top: Clay panel in bas-relief. 74 × 77 cm. Provenance, Palace of Agadja. Dahomey.—This naturalistic panel represents an allegory concerning the king Agonglo (1789–97). One explanation is that it is based on the saying, 'I am the pineapple against whom the lightning can do nothing'; signifying the pineapple growing beneath the palm which does not suffer even when the tree is struck by lightning. Another is that it refers to a proverb, 'The sheep do not eat the leaves of the palm tree'. The stylized representation of the palm tree shows a good sense of design.

Bottom: Clay panel in bas-relief. 73 × 75 cm. Provenance, Palace of Agadja. Dahomey.—This panel represents the conquest of the coastal territories and first contact with the people of Dahomey by the Europeans, who are represented by a seated Portuguese priest. The Dahomeans believed the Europeans could not walk as they always saw them seated. An extremely well designed panel.

PLATE VI. WALL DECORATION
Moulded decoration on the outer walls of houses in Bida, Northern Nigeria.—In this example the motifs are contained within panels. Imported plates and dish covers are embedded in various places to give colour to the design.

PLATE VII. WALL DECORATION

Moulded decoration on the outer wall of a house in Kano, Northern Nigeria.—The decoration covers the whole surface of the wall with no repetition of form except round the windows. Rich traders cover the front of their buildings with geometrical designs moulded with mud and covered with native cement from the dye pits, or with real cement. External decoration of this type is said to be of comparatively recent development, as until the end of the last century such patterns were only made on the interior walls of the houses.

PLATE VIII. WALL DECORATION

Top: Part of the vaulted ceiling in the reception room of the Emir of Kano. N. Nigeria.—Large coloured china plates are set in the mud at the intersection of the arches, like a boss. Interior decoration of this type preceded the exterior ornamentation of mud buildings at Kano. It was the work of women. See *Nigeria* 23. 1946.

Bottom: Coloured decoration on interior wall of hut. Hima tribe. Uganda.—Each of these motifs is symbolic, although the knowledge of the meanings of the symbols is fast dying out. Of these the following meanings are recorded. Top row reading from left to right: 1. The veil of modesty. (A veil made of strands of beads worn to conceal the face of a woman who worships the *Bacwezi* and who is possessed by a spirit. Her eyes must not be seen while she is in this condition.) 2. Two arrows. 3. Moons, the planet Venus, and smaller stars. 4. (*In centre to right of the veil*). A crowd of warriors in formation. 5. ?. Bottom row: 6. 'The Patterns'. (These are put on men's arms when the rains break and the herds return from the search for water.) 7. ?. 8. A spiral hair style.

See Sekintu and Wachsmann, 'Wall Patterns in Hima Huts'. Uganda Museum, 1956.

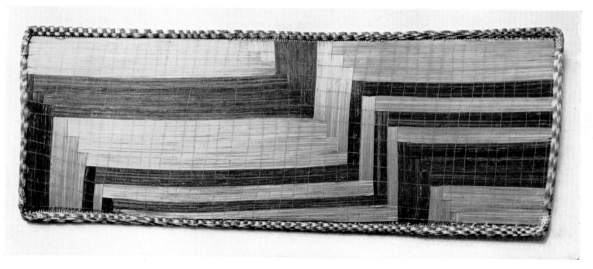

PLATE IX. WALL DECORATION

Top: Exterior wall of woven matting. Bushongo tribe. Belgian Congo.—Both interior and exterior walls of Bushongo huts may be covered with matting patterned in traditional design.

Bottom: Screen in sewn matting. 120 cm. × 45 cm. Tusi tribe. Ruanda Urundi. Collection, Margaret Trowell.—A screen used to partition off the platform on which the milk pots stand in a Tusi hut.

PLATE X. PATTERNS ON MATS AND SCREENS

Top: Woven mat. 175 cm. × 110 cm. Provenance, Boma. Sundi tribe. Lower Congo. Musée Royal du Congo Belge.—This woven mat has a warp of two colours, red-brown and purple, and is woven with a white weft which is floated behind and before a number of the warp strands so that the pattern can be seen in reverse on the other side of the mat.

Bottom: Sewn mat. Bushongo tribe. Central Congo. Musée Royal du Congo Belge.—This mat is made with a set of broad flat splints which are closely over-sewn with black and natural coloured raffia after the manner of coiled basketry. The reverse face can be seen rolled on the right hand side of the photograph. The motifs used are commonly found in Bushongo embroidery.

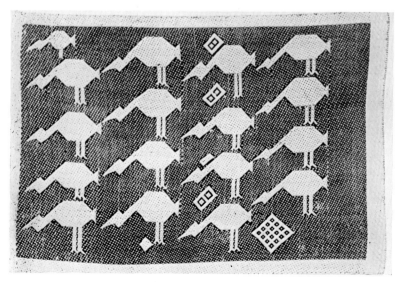

PLATE XI. PATTERNS ON MATS AND SCREENS

Top: Woven mat. Provenance, Cataracts. Lower Congo. Musée Royal du Congo Belge.—The weaving technique is clearly shown in this mat. The design appears in reverse, in white on black, on the other side of the work.

Centre: Woven mat. 171 cm. × 111 cm. Provenance, Cataracts. Lower Congo. Musée Royal du Congo Belge.—Technically this mat is woven in the same way as the one shown above. The human and geometrical motifs are more ambitious and better arranged.

Bottom: Woven mat. 159 cm. × 107 cm. Provenance, Cataracts. Lower Congo. Musée Royal du Congo Belge.—In the third example of this type of mat the bird motifs are used to make a definite repeat pattern.

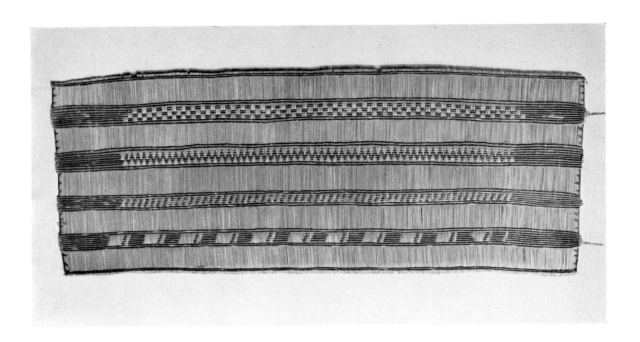

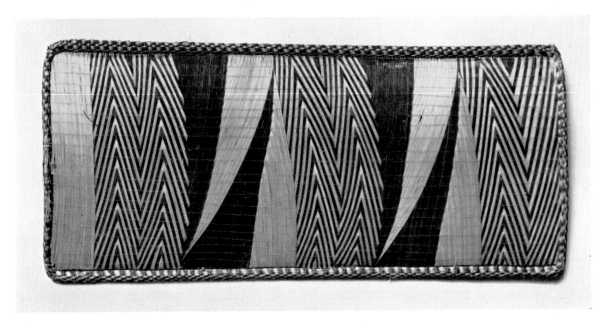

PLATE XII. PATTERNS ON MATS AND SCREENS

Top: Sewn mat. 108 cm. × 41 cm. Provenance, Mbarara. Hima tribe. Uganda. Uganda Museum.—The fine splints of the ground are sewn together by the black pattern work. This pattern is raised by a fibre cord under the stitching. The mat is worn draped over the head by an unmarried girl when leaving her own compound.

Bottom: Screen in sewn matting. 45 cm. × 95 cm. Tusi tribe. Ruanda Urundi. Collection, Margaret Trowell.—On top of a groundwork woven of flat splints the patterned covering of long black and natural coloured reeds is arranged and sewn in place with rows of fine fibre, the stitching being kept as invisible as possible. These screens are set up round the platform on which the sacred milk pots stand.

PLATE XIII. PATTERNS ON MATS AND SCREENS

Top: Set of patterns used in the making of plaited mats. Zanzibar. Collection, Margaret Trowell.—These mats are plaited in many bright colours, usually only one colour is used at a time combined with natural palm leaf. When the mats are made up the pattern motif forms an all-over pattern, or the decorated bands may be separated from each other by plain strips, or rows of different pattern may be sewn close together. The names of the patterns shown in this illustration are as follows (*reading from the left*): 1. The Window. 2. The Bat's Wing. 3. The Hen's Eye. 4. The Cow's Eye. 5. The Diamond. 6. The Cross. 7. The Heart. 8. The Snake. 9. The Bowtie. 10. The Lion's Foot. 11. The Fish Trap. 12. The Fish. 13. The Dog's Ribs. 14. The Starfish.

Bottom: Set of patterns used in the making of plaited mats. Nubian women in Uganda. Collection, Margaret Trowell.— Nubian women plait their mats with a double thickness of strands, working with one colour on top of another so that when the end of a plaited line is reached and the strands bent back to be worked in the opposite direction the colours are reversed. They rely on colour changes for the interest of their pattern rather than on the use of complicated motifs as can be seen from the sampler above, which contains seven variations on one basic pattern.

PLATE XIV. PATTERNS ON MATS AND SCREENS

Top: Prayer mat. Peace Memorial Museum. Zanzibar.—A very beautiful example of plaited matting. The broad strips of complicated pattern would entail a large number of strands in the plaiting.

Bottom: Prayer mat. Peace Memorial Museum. Zanzibar.—In this mat what appear to be Arabic characters have been used in certain stripes; their irregular form would involve very skilful arrangement of the plaited strands.

PLATE XV. TEXTILE DESIGN

Top: Woven cloths. Yoruba tribe. Nigeria.—These cloths are made of narrow strips sewn together.
Bottom left: Raffia woven mat. 69 cm. × 55 cm. No recorded provenance or documentation. Belgian Congo. Musée Royal du Congo Belge.—In both warp and weft black and white strands alternate, and are so woven as to give alternate vertically and horizontally striped squares; the squares themselves being in plain weave.
Bottom right: Silk Kente cloth. Nsuta. Ashanti tribe. Ghana. From the Beving Collection, late 19th or early 20th century. British Museum.— The cloth is made by sewing together a number of strips woven on a narrow loom. The common form of loom used by the Ashanti has two pairs of healds, one to weave the plain web and the second to produce the decorative pattern which is woven with additional wefts. These silk cloths are a blaze of brilliant colour—red, blue, green, black, white and yellow. Each design motif is named and has its own special significance.

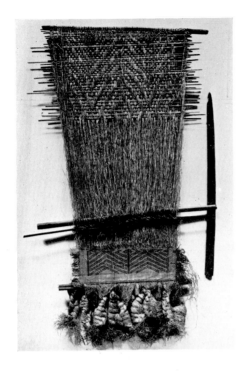

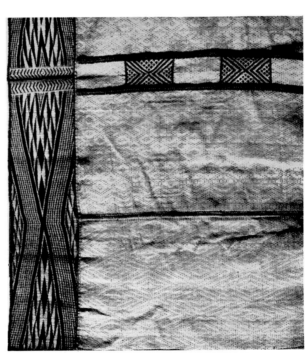

PLATE XVI. TEXTILE DESIGN

Top left: Loom for weaving raffia palm leaf fibre. Tetela tribe. Central Belgian Congo. British Museum.—The loom is provided with a heddle, shed stick and beater. (The latter standing vertically on the right in the photograph.) At the top of the warp can be seen some three dozen pattern rods threaded through the warp so that each may form the required shed for its place in the pattern. These rods are small in diameter and could not be used as shed sticks to open the sheds, so that it is probable that the weaver uses his fingers to trace down the openings to the weaving level of the textile. As each pattern rod is used it is withdrawn, and if required again replaced above the other rods in readiness for a repetition of the pattern.

Top right: Detail of woven raffia cloth. Kela tribe. Central Belgian Congo. Pitt Rivers Museum, University of Oxford.

Bottom left: Woven raffia cloth. 70 cm. × 38 cm. Jonga tribe. Central Belgian Congo. Musée Royal du Congo Belge.— This mat is typical of the type woven on the loom shown in *Top left*. The light design in red-brown and black on a natural ground gives a very delicate effect.

Bottom right: Woven raffia cloth. Mbun tribe. Kwango. Belgian Congo. British Museum.—The pattern of the side panels of this cloth shows the type of geometrical motifs which are used in African weaving. The central panels show two variations on the lozenge motif in diaper or one colour weaving.

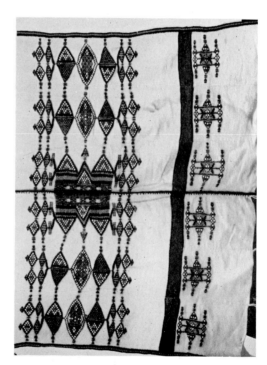

PLATE XVII. TEXTILE DESIGN

Top left: Woven cloth. Keta cloth. Ghana. Beving Collection, late 19th or early 20th century. British Museum.—This cloth is made up of two types of alternating narrow bands sewn together. The first is red crossed with dark blue, giving a purple effect, the second white. Both are decorated with stripes and other motifs including a bird and a fish woven with extra wefts.

Top right: Woven cloth. Probably Southern Nigeria. Beving Collection, late 19th or early 20th century. British Museum. A cloth with alternating horizontal panels of stripes and various geometrical motifs in brocaded weave.

Bottom left: Woven cloth. Provenance, Bamenda. Cameroons. Probably from the Niger Delta region. Pitt Rivers Museum, University of Oxford.—A white warp with a thin blue thread every quarter of an inch. The pattern is brocaded, one motif is possibly a human symbol.

Bottom right: Detail of woven pattern on a blanket. Provenance, Lake Cuallando. Debo. Fulani tribe. French Sudan. Pitt Rivers Museum, University of Oxford.

2

PLATE XVIII. TEXTILE DESIGN

Woven cloth. Upper Senegal. Beving Collection, late 19th or early 20th century. British Museum.—A cloth
made from strips woven on a narrow loom with a very intricate design in dark blue and white.

PLATE XIX. TEXTILE DESIGN

Top left: Detail of pile cloth. Probably the old kingdom of Congo. Lower Congo. Pitt Rivers Museum, University of Oxford.—The darker parts of this design are formed by embroidering the areas with finely shredded, dyed raffia, every stitch of which is cut off close to the surface giving a texture like pile velvet. In the lighter spaces in between the pattern is made by diaper weaving in natural coloured raffia. This cloth was collected before 1883, and is thought to have formed part of the Tradescant collection, acquired in the seventeenth century.

Top centre: Detail of pile cloth. Provenance, Boma. Probably Bushongo tribe. Central Belgian Congo. Pitt Rivers Museum, University of Oxford.—In this cloth, collected about 1910, the pile areas are made in the same way as in *top left*, while parts in between are embroidered with four rows of oversewing.

Top right: Detail of border of pile cloth. Provenance, Kasai area. Central Belgian Congo. Pitt Rivers Museum, University of Oxford.—This strip of border pattern shows the Pile cloth technique developed into the making of little bobbles of raffia; the bobbles are light and dark blue, warm yellow ochre and beige, while the ground is Indian red. The whole effect is that of a mosaic.

Bottom: Raffia pile cloth. 21 cm. × 32 cm. Bushongo tribe. Central Belgian Congo. British Museum.—Typical patterns found in Bushongo embroidery.

1 *Mbolo*   2 *Mbolo*   3 *Mbolo* and *Namba*

4 *Mosala Baba*

7   6 *Mbolo*   5 *Namba*

PLATE XX. TEXTILE DESIGN

Top: Embroidered raffia cloth. Bambala sub-tribe of Bushongo tribe. Belgian Congo. British Museum.—This cloth dates back to the 18th century. The cloths embroidered in this way, in contrast to those used for pile cloth, were very finely woven. Much of the work was done in a complicated form of chain stitch; and parts appear to be embroidered in a type of drawn thread work.

Bottom: Embroidered raffia cloths. Bambala sub-tribe of Bushongo tribe. Belgian Congo. British Museum.—Further examples of 18th century cloths. These cloths show examples of a number of traditional patterns; the names of those referred to in the chapter on geometrical pattern motifs are given above.

1

2

Mamanye 3

4

Mayulu 5

6

7 *Ikunji*

8 *Mbolo*

9 *Mongo*

10 *Lori*
Yongolo

11 *Nemo*
Kanya

12 *Mosala*
Baba

PLATE XXI. TEXTILE DESIGN

Embroidered Cloths. Bambala sub-tribe. Bushongo tribe. Belgian Congo. British Museum.—Further examples of
traditional pattern motifs are shown here.

PLATE XXII. TEXTILE DESIGN

Top: Detail of embroidered robe. Provenance, Bida. Nupe tribe. Northern Nigeria. British Museum.—Part of a brilliantly
embroidered robe in gold and silver thread on a scarlet ground.

Bottom: Detail of embroidered robe. Hausa tribe. Northern Nigeria. Pitt Rivers Museum, University of Oxford
White embroidery on a blue ground.

PLATE XXIII. TEXTILE DESIGN

Top: Embroidered robe. Bamenda. Cameroons. Pitt Rivers Museum, University of Oxford. Embroidered in red, white and yellow cotton on a blue ground.

Bottom: Detail of embroidery on very wide trousers. Probably Hausa tribe. Nigeria. Pitt Rivers Museum, University of Oxford.—The 'cuff' of the garment is woven in red and buff stripes. The main part of the garment is cream embroidered chiefly in red with some blue, green, black and yellow. A far less formally arranged design than the one shown above.

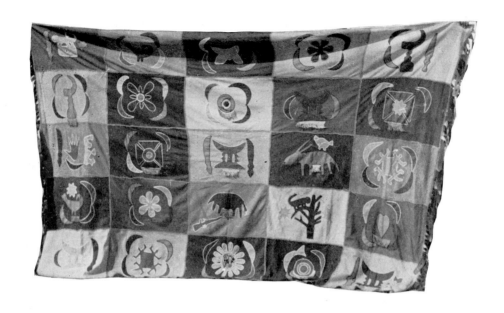

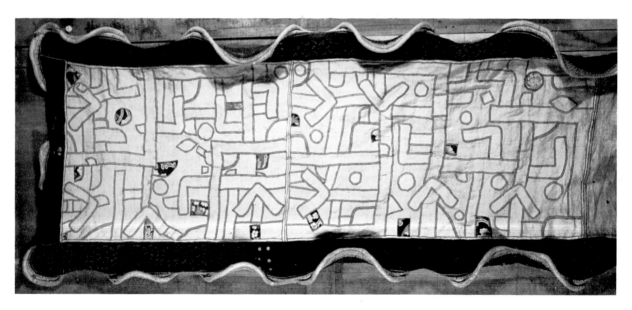

PLATE XXIV. TEXTILE DESIGN

Top: Appliqué cloth. An appliqué cloth of the Gyamanhene. Ashanti tribe. Ghana.
Bottom: Appliqué cloth. 161 cm. × 69 cm. Bushongo tribe. Belgian Congo. Musée Royal du Congo Belge.—A palm cloth with appliqué strips of the same material together with a few patches of imported prints and a border of pile cloth with insets of imported red fabric.

PLATE XXV. TEXTILE DESIGN

Appliqué cloth. 177 cm. × 107 cm. Provenance, Abomey. Fon tribe. Dahomey. Musée de l'Homme.—In this cloth the king is given an animal form symbolizing strength and is represented as far larger than his retainers and his enemies. Note that the retainers and the enemies have been cut from two patterns.

PLATE XXVI. TEXTILE DESIGN

Top: Painted barkcloth. 135 cm. × 130 cm. Mangbetu Tribe. Uele. Belgian Congo. Musée Royal du Congo Belge.
A very freely drawn design with a curious H-like motif constantly repeated.

Bottom: Painted raffia cloth. 161 cm. × 99 cm. No provenance. Belgian Congo. Musée Royal du Congo Belge.—An interesting design, or series of designs, are painted on the natural coloured ground of this raffia cloth. The solid forms are outlined with a narrow border of dark grey and filled in with brown. Dark grey spots cover the lighter areas.

PLATE XXVII. TEXTILE DESIGN

Painted cloth. No provenance. Ghana. British Museum.—The cloth is of white drill, roughly painted with an elaborate design very reminiscent of the embroidered robes of the northern tribes of the West Coast. The solid forms are outlined in black and filled in with green, red and yellow. Much of the background is covered with black 'writing' patterns which would appear to have been done with a pen.

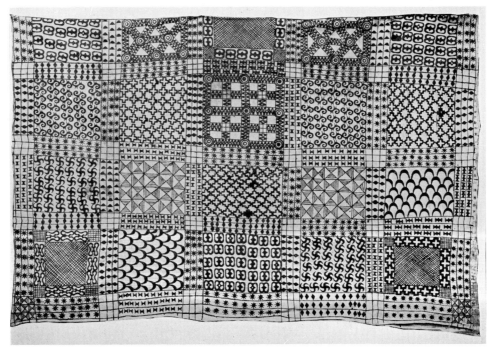

PLATE XXVIII. TEXTILE DESIGN

Top: Stamps used in printing *Adinkira* cloths. Ashanti tribe. Ghana. British Museum.—The stamps are cut from small pieces of calabash. Each one is named and has some symbolic meaning, some are the prerogative of the Royal household alone.

Bottom: Adinkira cloth. Provenance, Bonwere near Kumasi. Ashanti tribe. Ghana. British Museum.—This cloth, collected before 1934, shows the use of the stamp patterns illustrated above. The different motifs are usually printed in small panels either fitting closely together or separated by narrow bands of pattern as in this illustration.

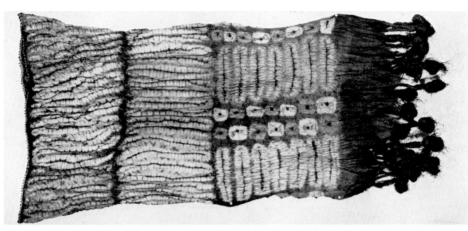

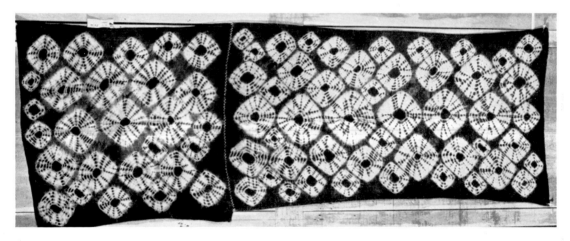

PLATE XXIX. TEXTILE DESIGN

Top left and top centre: Rope tie dyeing. Senegal. Beving Collection. British Museum.—The cloth is sewn and bound tightly together in parallel bands until it resembles a rope. When dyed and opened out the result is a series of horizontal undyed lines varying according to the closeness of the gathering and the thickness of the material.

Top right: Embroidered tie dyeing. Senegal. Beving Collection. British Museum.—The parts which are required to remain white have been closely oversewn in this specimen. The cloth will be dyed and the embroidery unpicked to get the desired effect.

Centre: Fine raffia cloth. Tie dyed. Ivory Coast. Musée de l'Homme.—This cloth appears to be plaited rather than woven, with the warp and weft threads on the bias as in certain basketry and mat-making techniques. It has been twice tied and dyed. The colouring is light yellow ochre, darker red ochre and black. The cloth has been left crimped and puckered.

Bottom: Tie dyed cloth. 150 cm. × 54 cm. Probably coastal region. Belgian Congo. Musée Royal du Congo Belge. A very accomplished piece of tie dyeing in black and natural raffia colours.

PLATE XXX. TEXTILE DESIGN

Top: Zinc stencil used in resist printing. Yoruba tribe. Nigeria.—The starch paste is painted on to the cloth through the stencil and allowed to dry. The cloth is then dyed, when the parts covered with the paste are unable to pick up the colour. The paste is finally removed from the cloth, leaving the design white against a dark ground.

Bottom: Resist printed cloth with stencil. Yoruba tribe. Nigeria.—Here a cloth is seen after dyeing.

PLATE XXXI. TEXTILE DESIGN

Top: Cloth printed by the Discharge method. Design showing red on a black ground. Bambara tribe. French Sudan. Musée de l'Homme.—The cloth having been dyed dark brown or black by soaking in a concoction made from the bark or leaves of certain trees, the design is then painted onto the cloth in a mud probably containing iron acetate. When this is dry the design is painted over a second time with a local soap made from ashes and vegetable oils containing a large amount of potash which acts as a mordant. It is finally painted yet a third time with the mud used in the first coat and thoroughly dried in the sun. The dye is thus chemically removed from the cloth in the painted area so that when the mud has been chipped and rubbed away the pattern stands out light against the dark ground. In this cloth and in the one shown below from the same area the designs have much in common. Both are treated as line drawings, with no solid masses; and certain motifs such as the zigzag line, lozenge shape, and the main form which divides up the broad border on the right of each cloth are used in both. There is sufficient variation in the different borders and panels to avoid monotony.

Bottom: Cloth printed by the Discharge method. Design showing red-brown on a black ground. Bambara tribe. French Sudan. Musée de l'Homme.—As above.

PLATE XXXII. TEXTILE DESIGN

Top: Cloth printed by the Discharge method. Bambara tribe. French Sudan. Musée de l'Homme.—Of the four cloths illustrated in this section this specimen is the most elaborate. Use is made of both solid light areas and fine line work, and the motif used in the squares of the central panel shows great variety of detail within its general uniformity. The same may be said for the other large motifs used in the borders.

Bottom: Cloth printed by the Discharge method. Bambara tribe. French Sudan. Musée de l'Homme.—The treatment of the pattern in this cloth is very similar to the one shown above.

PLATE XXXIII. ORNAMENTAL BASKETRY

Top left: Container of bark, covered with woven basketry. Height 31 cm. Diameter 27 cm. No provenance. Lower Congo. Musée Royal du Congo Belge.—In both this specimen and the next the container is made of bark, and the woven cover serves a decorative purpose only. Colouring black on a natural ground.

Top right: Container of bark, covered with woven basketry. No provenance. Lower Congo. Musée de l'Homme.—Colouring black on a natural ground.

Bottom left: Woven basket partly covered with more finely woven decorative outer layer. Height 21 cm. Diameter 37 cm. No provenance. Tanganyika. Manchester University Museum.—In this case the supporting layer is itself made of woven basketry. Colouring black on a natural ground.

Bottom right: Coiled basket. Barotseland. Northern Rhodesia. British Museum.—This large basket is decorated with geometrical pattern and birds and quadrupeds. Colouring black on a natural ground.

PLATE XXXIV. ORNAMENTAL BASKETRY

Top: Casket with decorative woven casing. No provenance. Probably Lower Congo. British Museum.—A very beautiful specimen of fine basketry which was acquired by the York Museum in 1827. Colouring black on a natural ground.

Bottom: Casket similar to the above. Length 43 cm. Height 32 cm. No provenance. Lower Congo. Musée Royal du Congo Belge.—Note the chevron pattern running round the body of the basket.

PLATE XXXV. ORNAMENTAL BASKETRY

Top: Three lidded baskets. Height I, 16 cm. (including lid); Height II, 13 cm. (including lid); Height III, 15 cm. (including lid). Tusi tribe. Ruanda Urundi. Musée Royal du Congo Belge.—The first and third of these baskets are woven, each in a different weave, the second is in coiled basketry.

Left centre: Woven stand for milk pot. Height 7 cm. Nyoro tribe. Uganda. Uganda Museum.—The pot stand is of coiled basketry with a very finely woven cover in red, black, and natural colour fibre.

Bottom left: Woven stand for milk pot. Height 9 cm. Nyoro tribe. Uganda. Uganda Museum.—The pattern in this pot stand is in natural, black and rich purple.

Bottom right: Two baskets. Height I, 17 cm.; Height II, 42 cm. (including lid); Tusi tribe. Ruanda Urundi. Musée Royal du Congo Belge.—The first of these baskets is woven, the second in coiled basketry. Both are patterned in black and natural colours.

All the small baskets and pot stands shown on these plates are of extremely fine and delicate workmanship.

PLATE XXXVI. ORNAMENTAL BASKETRY

Top left: Lidded basket. Provenance, Kwango area. Mbun tribe. Belgian Congo. British Museum.—The decorative interest of this basket lies in the use of different types of weave for the lid, the patterned band, and the base, with the raised bound bands dividing one area from another. The basket is all in natural colour.

Top right: Basket. Provenance, Kwango area. Mbun tribe. Belgian Congo. British Museum.—As in the previous basket, decorative interest is achieved by the method of construction alone, the vertical splints being held in place by a broad band of weaving round the centre which contrasts with the more open effect above and below and in the triangle left uncovered in the middle of the band.

Bottom left: Woven basket, with attached decorative band. Height 345 cm. Diameter at top 305 cm. Provenance, Kwango area. Mbala tribe. Belgian Congo. Musée Royal du Congo Belge.—The band attached to this basket serves the function of a handle, but has also been constructed with due regard to decorative values. Black design on natural colour ground.

Bottom right: Coiled basket on a wooden base. Provenance, Kasai. Bushongo tribe, Mbala sub-tribe. Belgian Congo. British Museum.—A royal basket said to contain the king's wisdom. The carved wooden base of this basket gives dignity to an otherwise not very outstanding design, although there is interest in the contrasting weight of the weaves in the pattern areas. Natural colour.

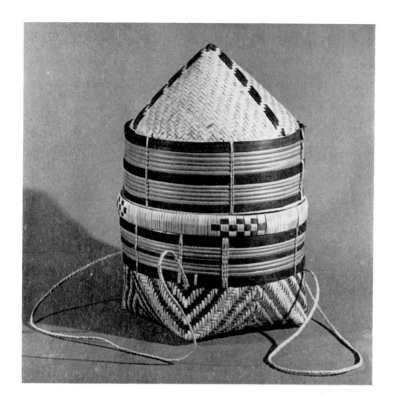

PLATE XXXVII. ORNAMENTAL BASKETRY

Top: Lidded basket. Height 36 cm. Diameter 24 cm. No provenance. Belgian Congo. Musée Royal du Congo Belge.
This is a most interesting small basket. The square based basket becomes circular at the top, and further interest is
added by the conical lid and the many bands of unwoven cane.
Bottom: Lidded basket. Height 35 cm. Diameter 38 cm. No provenance. Belgian Congo. Musée Royal du Congo Belge.
The decorative portion of this basket is made of two bands of blackened wood carved in very low relief and attached
to the base and lid.

PLATE XXXVIII. BEADWORK

Top: Beaded bag. Yoruba tribe. Nigeria. British Museum.—The colouring of this bag, presented to the Governor of Lagos at the beginning of this century, is far richer and more varied than that of modern beadwork.

Bottom: Beaded boots. Yoruba tribe. Nigeria. British Museum.—These boots have an interesting history. According to tradition Oduduwa, the first ruler of Yoruba, gave beaded crowns to his sixteen sons, who became rulers of the different Yoruba states. (See Mellor, 'Bead Embroideries of Remo,' *Nigeria* 14, 1938, p. 154; also *Nigeria* 40, 1953, p. 306. (Photograph of beaded crown worn by the *Ataoja* of Oshogbo.)) The rulers of these states are alone entitled to wear such crowns, which had sixteen birds worked upon them. Besides the crowns were other articles of apparel, including boots. Later other chiefs pretended to equal rank, but their claims were disallowed. These boots, with eight stuffed beaded birds running up the front, belonged to a chief styled the Elepe of Epe, and were forfeited together with other articles when his claim was rejected. They were acquired by the British Museum in 1904.

PLATE XXXIX. BEADWORK

Top left: Beaded covering of a calabash. Bali tribe. Cameroons. British Museum.
Top right: Beaded belts. Bantu tribes. S. Africa. British Museum.
Bottom: Bead covered cases. Heights, I, 19 cm.; II, 10 cm.; III, 11 cm. Tusi tribe. Ruanda Urundi. Musée Royal du Congo Belge.

PLATE XL. THE DECORATION OF HIDES AND LEATHER

Top left: Shaven Pattern. Subi tribe. Tanganyika. King George V Museum. Dar es Salaam.—This pattern, on antelope skin, is formed by shaving away the hair in narrow lines.

Top right: Sheath with carved pattern. Height 45 cm. Bagam tribe. Cameroons. University Museum of Archaeology and Ethnology, Cambridge. This pattern is cut out in low relief and emphasized by rubbing in a white chalky substance.

Bottom left: Fan of cowhide with applied decoration. Benin. Nigeria. British Museum.—These circular fans of cowhide, ornamented with appliqué in red flannel and thin yellow skin, are mentioned frequently by early writers on Benin, and are represented in the hands of the King's retainers on many of the bronze plaques. Motifs used in their decoration seem to range from various symbols to leaf, animal and human forms.

Bottom right: Fan of cowhide with applied decoration. Provenance, Ogbe. Southern Nigeria. British Museum. As *bottom left* above.

 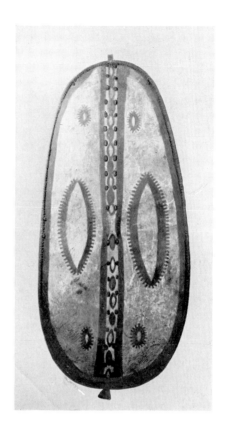

PLATE XLI. THE DECORATION OF HIDES AND LEATHER

Top left: Hide Shield. Height 110 cm. Masai tribe. Kenya. Manchester University Museum.—The decoration of these Masai shields is painted on in coloured earths.

Top right: Hide Shield. Height 125 cm. Masai tribe. Kenya. Manchester University Museum.

Bottom: Warriors with shields. Photograph by Hobley. Masai tribe. Kenya.

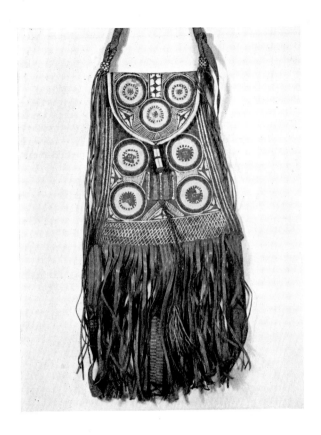
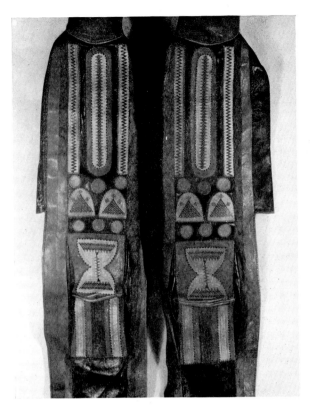

PLATE XLII. THE DECORATION OF HIDES AND LEATHER

Top left: Leather bag. Hausa tribe. N. Nigeria. British Museum.—The bag is decorated with appliqué and stitching in various contrasting colours, black, white and red on a dark tan background.

Top right: Part of pair of decorated leather boots. Provenance, Kano. Hausa tribe. N. Nigeria. British Museum. The decoration of these boots is similar in technique to that of the bag on *left*.

Bottom left: Bag of the Shango cult. Yoruba tribe. Nigeria. British Museum.—Although the appliqué decoration of this bag has not the same technical precision as that of the two examples from Northern Nigeria, it is artistically more interesting in its irregularity.

Bottom right: Leather bottle cover. Sierra Leone. British Museum.—The cover is of dark red leather, the pattern being made by sewing in fine strips of dried palm leaf such as are used in mat making.

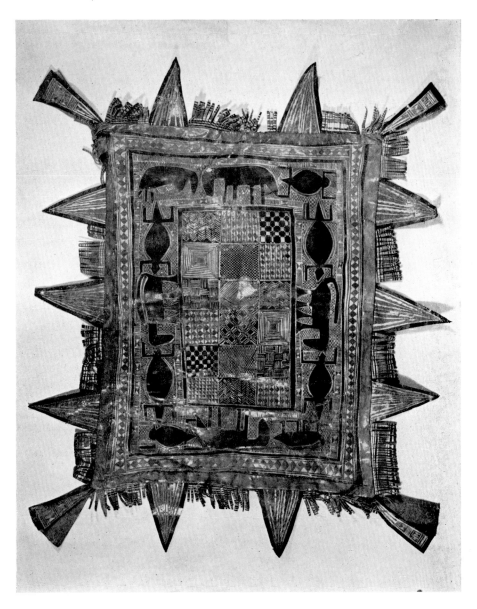

PLATE XLIII. THE DECORATION OF HIDES AND LEATHER

Cushion cover or mat. Provenance, Bida. Nupe tribe. N. Nigeria. British Museum.—This is a most interesting example of leatherwork, not only for its wealth of geometrical pattern motifs, but also for the use of animal, reptile and bird forms.

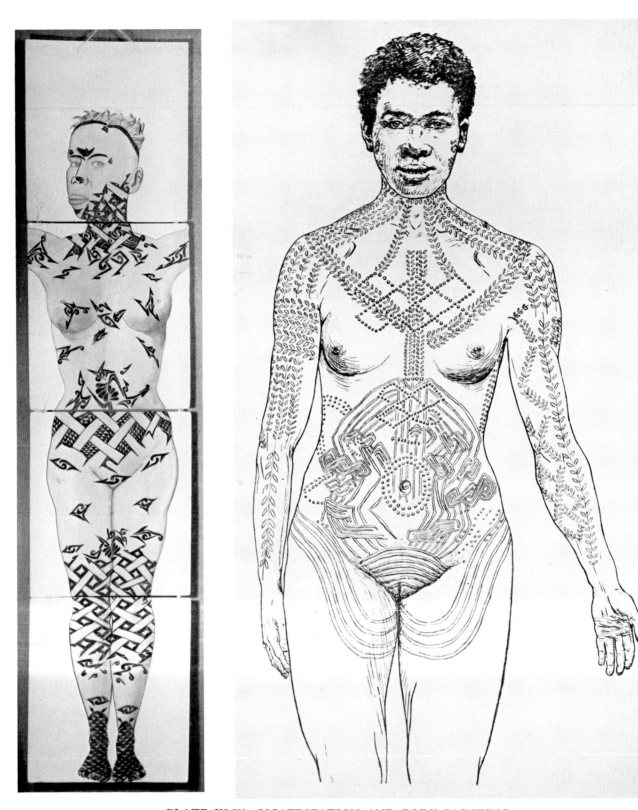

PLATE XLIV. CICATRIZATION AND BODY PAINTING

Left: Drawing showing correct placing of painted designs on the body. Tiv tribe. Nigeria.
Right: Drawing showing cicatrization patterns. Tetela (Sungu) tribe. Central Congo.

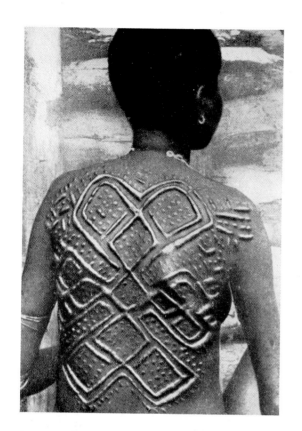
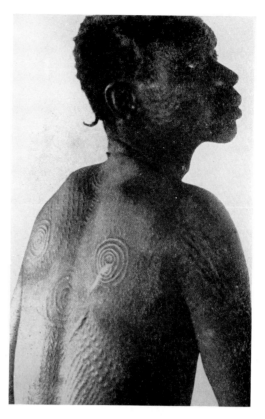
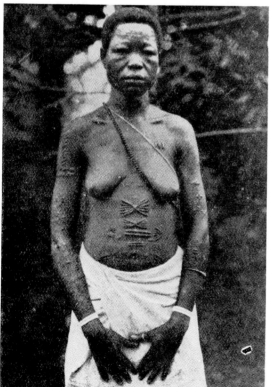

PLATE XLV. CICATRIZATION AND BODY PAINTING

Top left: Cicatrization pattern on woman's back. Mayombe tribe. Lower Congo.
Top right: Cicatrization pattern on woman's back. Nkutshu (Bankutu) tribe. Central Congo.
Bottom left: Cicatrization pattern on woman's abdomen. Tetela (Sungu) tribe. Central Congo.
Bottom right: Cicatrization pattern on the calves of the legs. Tiv Tribe. Nigeria.

PLATE XLVI. CALABASH PATTERNS

Top: Carved calabash. Provenance, Lagos. Southern Nigeria. Manchester University Museum.—This calabash has been scraped before being carved with geometrical pattern. Portions of the pattern were then left white, while others were stained. The background of the pattern has been very deeply cut away.

Bottom: Carved calabash. Provenance, Ilorin. Northern Yoruba tribe. British Museum.—Another well carved calabash in which the shadows cast by the angular cutting give an added interest to the pattern.

PLATE XLVII. CALABASH PATTERNS

Scraped calabash lids. Bariba tribe. Dahomey. Musée de l'Homme.—In these calabash lids the background has been scraped away leaving the pattern motifs raised. Further contrast is given by the engraving of small texture patterns on the animal motifs contained within the panels, while treating the horizontal bands of geometrical pattern more broadly. Of these calabashes from Dahomey, Griaule and Dieterlen say that the panels are often filled with representations of animals or geometric pattern, sometimes with representations of objects, much more rarely with men. Herskovits also emphasizes the rarity of anthropomorphic representation. He says that the decorations convey messages to the initiated as clearly as would writing to the literate, and they represent symbolic stories, proverbs, or very often love messages, for they are presented by young men to the girl of their fancy. These calabashes from Dahomey are some of the most beautiful examples of calabash decoration and are clearly appreciated locally for aesthetic reasons.

PLATE XLVIII. CALABASH PATTERNS

Top: Two views of a scraped calabash drinking vessel. Dahomey. Musée de l'Homme.—This very fine example of a Dahomean calabash is decorated by the scraping technique, the cuticle of the background portions having been scraped away. The use of human motifs is somewhat rare; a third panel shows a bicycle. Each panel is well designed in itself; and the balancing of the figured panels with the plain portions, together with the borders at the top and bottom, give a very pleasing total effect.

Bottom: Two scraped calabash drinking vessels. Bariba. Dahomey. Musée de l'Homme.—These two very similar calabashes are also decorated by the scraping away of the background. In the panels can be seen stylized birds and animals together with various other symbols.

PLATE XLIX. CALABASH PATTERNS

Top: Scraped calabash bowl. Dahomey. Musée de l'Homme.—As plates XLV and XLVI.

Bottom: Scraped and engraved Calabash. Foulbe tribe. Cameroons. Musée de l'Homme.—This calabash from the Cameroons shows an entirely different treatment from the previous examples. In this case a very rich effect has been achieved through cleaning away the surface colour of the calabash in a few areas only, and covering even these with fine texture patterns. The free geometric design is excellent, every part being covered with interesting motifs skilfully combined into a satisfying whole.

PLATE L. CALABASH PATTERNS

Top: Decorated calabash. Probably Nupe tribe. N. Nigeria. British Museum.—A very fine example of calabash decoration. The texture pattern filling in certain portions of the design is reminiscent of basketwork.

Bottom: Three calabash lids. S.E. Nigeria. British Museum.—These calabash lids are excellent specimens of decoration by the scraping method. The groundwork is the natural deep sienna colour of the gourd, but the patterned portions have had the top layer of cuticle scraped off, leaving a white surface, which has then been stained in places with black.

PLATE LI. CALABASH PATTERNS

Top: Scorched and engraved calabash. Provenance, Bauchi town. Hausa tribe. N. Nigeria. British Museum.—This calabash would appear to have first been scraped, and then engraved by burning with an iron point. The wide black band has been scorched with a flatter tool.

Bottom left: Scorched calabash. Height 25 cm. Provenance, Kigezi. Kiga tribe. Uganda. Uganda Museum.—The black pattern is obtained by scorching with the blade of a knife or similar piece of metal. The curvilinear pattern is typical of the work of this tribe, but rare in many parts of Africa.

Bottom right: Scorched and engraved Calabash ladle. Probably Ibibio tribe. S. Nigeria. British Museum.—The flowing leaf pattern is typical of south-eastern Nigeria; see also Plate LXI.

PLATE LII. CALABASH PATTERNS

Top left: Calabash bottle engraved with a hot point. Hausa tribe. N. Nigeria. British Museum.—Although roughly executed the filling in of the panels is good.

Top right: Engraved calabash bottle. Kamba tribe. Kenya. British Museum.—This very decorative specimen is richly engraved with a very fine line. The engraved lines have been filled with some black material.

Bottom left: Engraved calabash bottle. Height 25 cm. Probably Lower Congo. Musée Royal du Congo Belge.—Loosely arranged and drawn, the drummers and dancers here make an interesting design.

Bottom right: Engraved calabash bottle. Tribe and provenance Unknown. British Museum.—The figures on this specimen are scattered about in a haphazard fashion. The chief interest in the work lies in the technique. The background has first been cut away leaving the figures in very low relief, and then engraved with groups of parallel lines; the figures are blackened by scorching.

PLATE LIII. CALABASH PATTERNS

Top left and right: Calabashes decorated with extraneous materials. Provenances unknown. (*Top left*) West Africa. British Museum.—The body of the calabash is the natural burnt sienna colour, with bands of black scorched pattern. These have been outlined with small white beads pressed into the skin of the gourd. (*Top right*) S. Africa.—In this case the pattern is stitched on to the calabash in brass and steel wire.

Bottom: Carved coconut. Provenance, Benin. Nigeria. British Museum.—This work in low relief much more resembles wood carving from the same area than calabash decoration.

PLATE LIV. DECORATION ON WOOD

Top left: Wooden drinking vessel. Height 19 cm. Bushongo tribe. Congo. Musée Royal du Congo Belge.—This specimen shows the use of a most interesting mixture of all-over texture pattern together with representational motifs in low relief, as the texture ornament is carried up on to the backs of the frogs.

Top right: Small wooden keg for gunpowder. Height 9 cm. Lower Congo. Musée Royal du Congo Belge.—In this example a geometrical motif of concentric squares has been cut so closely together that the pattern assumes a textural value.

Bottom left: Carved wooden cup. Provenance, Kasai. Bushongo tribe. Congo. American Museum of Natural History. The interlacing straps and the spaces between them are filled respectively with parallel and cross-hatched lines forming an interesting texture pattern covering the whole vessel.

Bottom right: Carved wooden cup. Bushongo tribe. Congo. American Museum of Natural History.—In this cup more use has been made of spacing between the pattern bands. The pattern motif round the body of the cup is *Mbolo*.

PLATE LV. DECORATION ON WOOD

Top: Carved pattern on a head rest. Provenance, Mashonaland. Southern Rhodesia. British Museum.—These small head rests from Southern Rhodesia and Nyasaland demonstrate the African's ability to fit his pattern to whatever shapes he is presented with by the object he is decorating.

Bottom left: Wooden powder keg. Provenance, Loudima Niadi. Kamba tribe. French Equatorial Africa. Musée de l'Homme.—The tall oval shapes fit the shape of the body, while the diamond pattern formed by engraved cross-hatching gives an interesting texture.

Bottom right: Wooden powder keg. No provenance. Lower Congo. Musée Royal du Congo Belge.—A more sophisticated and better executed example than *Bottom left*, but somewhat similar in design.

PLATE LVI. DECORATION ON WOOD.

Wooden vessel. Provenance, Kasai. Belgian Congo. American Museum of Natural History.—An exceptionally lovely specimen of African carved decoration. The vessel itself is beautifully proportioned, and the closely incised texture pattern of the neck contrasting with the sweeping curves of interlacing pattern on the body emphasizes both the proportion and the shape. This pattern is known to the Bushongo tribe as *Buina*, and appears on many wooden objects.

PLATE LVII. DECORATION ON WOOD

Carved wooden door panel in low relief. Height 142 cm. Baule tribe. Ivory Coast. Collection. Mr. Josef Muller.—
A simple but extremely beautiful design. Note the contrast between the plain adzed wood of the background, with the
slight suggestion of waves, and the patterning of the fishes' bodies and fins.

PLATE LVIII. DECORATION ON WOOD

Left: Carved door panel. About 300 cm. high. Provenance, Ikare. Kabba Province. Eastern Yoruba. Nigeria. British Museum.—A good example of Yoruba door panelling with bands of processional figures divided by smaller bands of pattern.

Centre: Carved door panel. Provenance, Ijebu. Yoruba. Nigeria. British Museum.—A highly stylized design of figures and snakes. Note the small zig-zag pattern which is carried across each form.

Right: Part of a carved door-frame. Arab. Zanzibar.—Arab and Swahili carving was being replaced by the work of local Indians in Zanzibar early in this century, but some still remained. 'The chief carving designs' (in Arab and Swahili carving) 'are three in number: lotus derivatives, the rosette, and a frankincense or date palm derivative. In every door-frame the designs on the two uprights spring at the foot from fish-like objects, figured, apparently, in a head-downwards position, which are generally represented with a conventional scale pattern surface, and immediately below these fish-like objects there are sometimes figured two or three wavy lines, suggestive of the Egyptian hieroglyphic for water.' See Barton, 'Zanzibar doors,' *Man.* June 1924, No. 63.

PLATE LIX. DECORATION ON WOOD

Top left: Carved wooden vessel. Height 22 cm. Bushongo tribe. Congo. Musée Royal du Congo Belge.—An elaborate and skilfully carved pattern.

Top right: Carved wooden drinking vessel. Height 19 cm. Bushongo tribe. Congo. Musée Royal du Congo Belge. —A rather unusual type of pattern covers this pot.

Bottom left: Carved wooden drum. Provenance, Kasai. Bushongo tribe. Congo. British Museum.—A very rich effect has been obtained here with the lattice work of knotted pattern known as *Namba* encasing the sun motif. The drum probably dates back to the reign of Bom Bosh, about the mid-seventeenth century.

Bottom right: Carved wooden drum. Provenance, Ijebu. Yoruba. Nigeria. British Museum.—The whole of this side of the drum is covered with a very stylized human figure; the legs becoming fish forms and turning upwards to be held by the hands; and hornlike appendages hanging down on either side of the large head, very common motifs in both wood and ivory carving from this district. Texture patterns cover most of the surface of the designs.

PLATE LX. DECORATION ON WOOD

Wooden milk pot. Height 34 cm. French Somaliland. Musée de l'Homme.—This milk pot is decorated in black and white work. The surface of the vessel is first blackened and then carved in low relief.

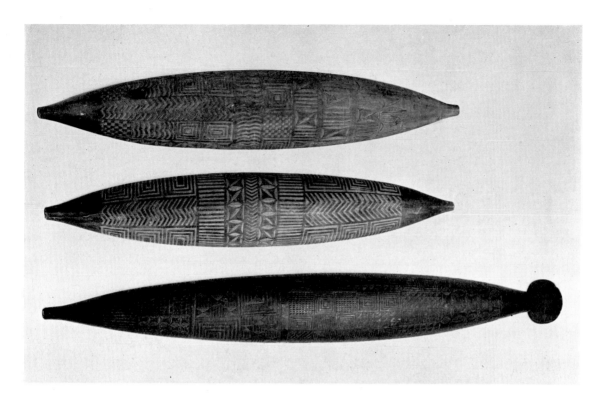

PLATE LXI. DECORATION ON WOOD

Top: Model canoes. Sierra Leone. British Museum. The two top canoes shown here are in black and white work.
Bottom: Wooden fans. Ibibio tribe. S. Nigeria. British Museum.—These fans are decorated with patterns engraved with a hot metal point and blackened with a small flat metal tool. The flowing leaf-like forms are typical of the work of Eastern Nigeria.

PLATE LXII. DECORATION ON WOOD

Top: Two wooden shields and a decorated board used in dancing. Kikuyu tribe. Kenya. British Museum.—The shield
on the left and dancing board on the right are in black and white work, while the central shield is painted.
Bottom: Three wooden shields. 1. Twa tribe. Ruanda Urundi. 2. Ganda tribe. Uganda. 3. Tusi tribe. Ruanda Urundi.
British Museum. Examples of painted shields.

PLATE LXIII. DECORATION ON WOOD

Top: Stool with wire decoration. Kamba tribe. Kenya. Manchester University Museum.—The wire used in decorating this stool has first been wound tightly round another piece of metal to twist it into a spiral, and then hammered into the surface of the wood which has been softened with oil.

Bottom: Stools with wire decoration. Kamba tribe. Kenya. British Museum.—These stools are decorated in a similar fashion to the one above.

PLATE LXIV. ORNAMENTAL IVORY CARVING

Top: Ivory bracelet. Provenance, Benin City. Probably Yoruba, Nigeria. British Museum.—The bracelet is carved in two layers from one piece of tusk; the inner one, which is decorated with drilled holes slipping round easily, but prevented from coming apart by two pegs of which the small bird at the centre base is one. The central figure of the king is the traditional representation carved symmetrically in a frontal position with legs developed into fish forms; while the other figures are freely and naturalistically carved.

Bottom: Ivory bowl. Height 12 cm. Provenance, Owo district. Yoruba, Nigeria. British Museum.—The figures on this carving are fairly naturalistic and clearly arranged in solid masses.

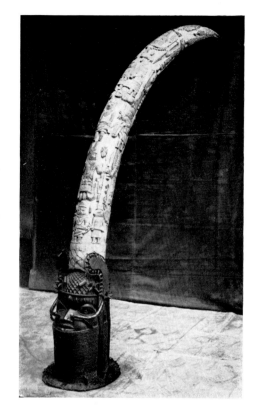

PLATE LXV. ORNAMENTAL IVORY CARVING

Top left: Ivory tusks. Provenance, Loango. Lower Congo. British Museum.—In these two tusks from the Lower Congo the figures are far more naturalistic and arranged in a more orderly way than those on the Benin tusk shown on the right.

Top right: Carved ivory tusk. Provenance, Benin. Nigeria. British Museum.—This tusk is typical of those found on the royal altars and burial chambers. A broad band of interlacing pattern surrounds the base, while above figures and emblems are arranged over the whole surface in an irregular but well-balanced manner. Both the stylized and naturalistic trends of Yoruba and Benin work can be seen here, the former in the representation of kings with supporting attendants on either side, and legs developed into fish forms; and the latter in the three quarter view of a horseman half-way up the tusk on the right-hand side.

Bottom, left and right: Carved ivory jug. Height 20 cm. Provenance, Benin. Probably Yoruba tribe. Nigeria. British Museum.—The metal rim and base, together with the wooden handle, are later additions. Of the motifs shown in these two views of the jug, the two headed bird in the upper stage of *bottom left*, the mud fish in the upper stage of *bottom right* and the elephant head in the lower stage of *bottom right* with the trunk divided into two arms of which the hands hold branches, are common symbols in Benin work, while the grazing antelope in the lower stage of *bottom left* is completely naturalistic.

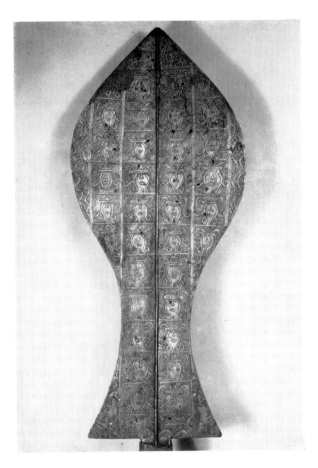

PLATE LXVI. DECORATIVE METAL WORK

Top left: Armlet in repoussé work. Provenance, Benin. Nigeria. British Museum.

Top right: Brass *Ebere*. Provenance, Benin. Nigeria. British Museum.—These objects, usually about 90 cms. long, were not weapons but ornaments carried by Court officials. The one illustrated here is decorated with a repetition of a very stylized human head, probably meant to represent a Portuguese.

Bottom, left and right: Plaques decorated in repoussé. Diameter in each case 27 cm. Provenance, Benin. Nigeria. British Museum.—Thin brass sheeting decorated in repoussé. Probably late 19th century. Note that the figure *bottom left* is represented in the same way as that on the armlet (*top left*), while the head of the figure in *bottom right* is a rather more naturalistic version of that in *top right*.

PLATE LXVII. DECORATIVE METAL WORK

Top left: Cast brass *kuduo* or ceremonial vessel. Ashanti tribe. Ghana. British Museum.—In cast work the pattern is engraved in the wax.
The openwork stand has been brazed on to the vessel.

Top right: Repoussé brass vessel. Provenance, Attabubu, north of Kumasi. Ashanti tribe. Ghana. British Museum.—The design on this
vessel is punched into the brass and filled with a white substance.

Bottom left: Beaten gold Soul-Bearer's disc. Ashanti tribe. Ghana. British Museum.

Bottom right: Cast gold Soul-Bearer's disc. Ashanti tribe. Ghana. British Museum.

PLATE LXVIII. DECORATIVE METAL WORK

Top: Bronze plaque. Benin. Nigeria. British Museum.—This seventeenth-century plaque shows guards and attendants at the gate of the Oba's palace. Note the variety of pattern on the walls of the building, and attendants' shields and clothing.

Bottom: Bronze plaque. Benin. Nigeria. British Museum.—In this group representing the Oba with his attendants a great deal of naturalism has been attained while the convention of strict frontality is still adhered to. This combination of formal, symmetrical grouping, together with free rhythm, gives the work the quality of good design. The texture patterning of the clothing is even more interesting than in the previous example. The work is probably eighteenth century.

PLATE LXIX. DECORATIVE METAL WORK

Top: Bronze plaque. Benin. Nigeria. British Museum.—This plaque is probably seventeenth-century work, and one of the earliest styles known. It is considerably more rigid than those shown in the previous plate, both in the grouping of the figures and the pattern work.

Bottom: Bronze plaque. 48 cm. Benin. Nigeria.—The Portuguese shown in this plaque is most realistically represented. Special notice should be taken of his head, with its flowing hair and beard, long narrow nose and mouth, and helmet. Reference should then be made to the stylized types of Portuguese heads shown in Plate LXVI *bottom right and top right*. Here the characteristic features mentioned above are reduced to a decorative symbol.

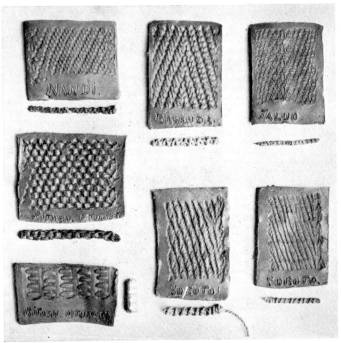

PLATE LXX. POTTERY DESIGN

Top left: Red earthenware water pot with impressed pattern. Height 35 cm. Nyoro tribe. Uganda. Uganda Museum.—A very fine example of a roulette impressed pot.

Top right: Impressed pattern on a clay head. Tribe unknown. American Museum of Natural History.—No details are available of this work, but the decoration of the face with finger nail indentations is exceedingly interesting.

Bottom left: Buff earthenware pot decorated with impressed pattern. Height 25 cm. Budu tribe. Uele. Belgian Congo. Musée Royal du Congo Belge.—A biscuit coloured pot with broad bands of roulette impressed pattern outlined with an incised line. The interest of the design lies in the arrangement of the pattern bands which is most unusual.

Bottom right: Examples of roulettes with specimen impressions. East African tribes. British Museum.—The roulettes are placed below the clay slabs showing their impressions. That in the bottom left-hand corner is of wood, the rest of reed or fibre.

PLATE LXXI. POTTERY DESIGN

Top left: Cup or pot with incised pattern. Height 10 cm. No documentation. Nyoro or Hima tribe. Uganda. Uganda Museum.—
The groundwork of this pot is burnished black, and the finely incised pattern has been filled with white chalk or crushed shell.
Top right: Vase with incised and impressed decoration. Height 36 cm. Teke tribe. Lower Congo. Musée Royal du Congo Belge.
—The ground of the pot is a very light buff, and parts of the incised decoration have been filled in with a dark red. Parts
of the pattern are incised and the broken lines may be impressed with a metal ring.
Bottom left: Earthenware pot partially burnished, with incised and impressed decoration. Lower Congo. Musée Royal du
Congo Belge.—The neck and lower portion of this pot have been burnished, while the matt surface of the upper half of the
body is patterned. The horizontal grooves are deeply incised, and given interest by irregular verticals. The bands formed
between the grooves are further enriched by the lightly impressed pattern.
Bottom right: Earthenware pot, burnished and incised. Mayombe tribe. Lower Congo. Musée Royal du Congo Belge.—The
neck and upper part of the body, except for the patterned areas, are burnished a deep red, the lower part of the body has a
matt, unburnished surface. The pattern is incised with a deep, sharply cut line.

PLATE LXXII. POTTERY DESIGN

Top left: Black earthenware pot with moulded decoration. Provenance. Fumbam. Bamum tribe. Cameroons. Musée de l'Homme.—The stylized figures round the body of this pot, with the arms and legs forming a latticed pattern, are typical of the work of the Cameroons both in pottery and wood carving.

Top right and bottom: Three ceremonial vessels of black earthenware. Ashanti tribe. Ghana. British Museum.—The decoration of these pots combines moulded, impressed and incised work.

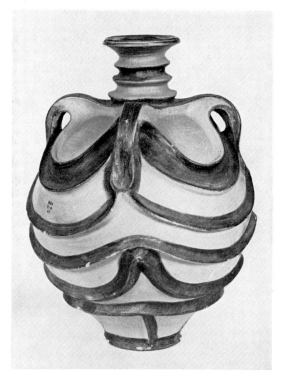

PLATE LXXIII. POTTERY DESIGN

Top left: Pot with moulded design. Provenance, Niger Delta. Ibo tribe. Nigeria. British Museum.—The lines of decoration of this pot have probably been pinched up while the clay was still soft, and then roughly impressed with a finger nail or piece of wood.

Top right: Pot with moulded decoration. Height 12 cm. No tribe or provenance. Congo. Musée Royal du Congo Belge. This very well designed little pot is in a very light coloured clay, the raised pattern is simple and balanced.

Bottom left: Water pot with moulded design. Provenance, Inye. Northern Ibo tribe. Nigeria. British Museum.—The very large pot is covered with deep grooves and ridges which have been darkened with a slip of coloured clay.

Bottom right: Earthenware pot with moulded and impressed decoration. No documentation. Congo. Musée Royal du Congo Belge.—The grooved neck and deeply gouged modelling on the lower half of the body of the pot are very well balanced. The impressed pattern on the surface surrounding the gouged curves is probably made with a metal ring.

PLATE LXXIV. POTTERY DESIGN

Top left: Black earthenware pot with moulded decoration. No documentation. (?) Zulu. S. Africa. British Museum. The regular bands of moulded pellets on this pot form an interesting contrast to the more angular knobs used in the next illustration. Their arrangement on the body of the pot is excellent.

Top right: Earthenware pot with moulded decoration. No documentation. Congo. Musée Royal du Congo Belge.—The decoration on the upper half of the body of this pot is much more roughly executed than that of the previous one, but the total effect has a vitality which the other lacks.

Bottom left: Earthenware pot splashed with vegetable matter. Height 32 cm. Sundi or related tribe. Lower Congo. Musée Royal du Congo Belge.—Both from the point of view of its general shape and its decoration this is a very beautiful pot; the basic colour is a very pale pinkish biscuit, and the splashed pattern is in several shades of brown.

Bottom right: Painted earthenware pot. Height 30 cm. Provenance, Stanley Pool. Probably Teke tribe. Lower Congo. Musée Royal du Congo Belge.—The body of this pot is roughly painted in vandyke brown.

PLATE LXXV. POTTERY DESIGN

Earthenware pot. Ashanti tribe. Ghana. British Museum.—This ceremonial pot which was used to contain the wine poured upon the Golden Stool has a very fine design which can also be seen on the gold Soul-Bearer's disc illustrated in Plate LXVII (*bottom right*).

PLATE LXXVI. POTTERY DESIGN

Top left: Encased pot. Height 20 cm. Angba tribe. Uele. Congo. Musée Royal du Congo Belge.—The neck and base of this pot are encased with a woven fibre covering which is continued as a handle. The body has a decorative binding of strips of cane framing small panels which show roulette decoration.

Top right: Encased pot. Height 32 cm. Bali tribe. Uele. Congo. Musée Royal du Congo Belge.—The whole pot is encased in a fibre covering which in itself is a beautifully decorative piece of work.

Bottom, left and right: Cow dung pots. Provenance, Kadero. Nuba Hills. Kordofan. British Museum.—These two bowls which are made of cow dung are painted in white and red earth on the dark brown surface and are excellent examples of rich, freely painted design on pottery.